CISM COURSES AND LECTURES

The series presents lecture notes, monographs, edited works and proceedings in the field of Mechanics, Engineering, Computer Science and Applied Mathematics.
Purpose of the series is to make known in the international scientific and technical community results obtained in some of the activities organized by CISM, the International Centre for Mechanical Sciences.

INTERNATIONAL CENTRE FOR MECHANICAL SCIENCES

COURSES AND LECTURES - No. 418

MAGNETOHYDRODYNAMICS

EDITED BY

PETER A. DAVIDSON
UNIVERSITY OF CAMBRIDGE

ANDRE THESS
UNIVERSITY OF ILMENAU

Lecture notes of the 1999 summer school organised by IUTAM and HYDROMAG

Springer-Verlag Wien GmbH

This volume contains 62 illustrations

Originally published by Springer-Verlag Wien New York in 2002
SPIN 10877352

In order to make this volume available as economically and as rapidly as possible the authors' typescripts have been reproduced in their original forms. This method unfortunately has its typographical limitations but it is hoped that they in no way distract the reader.

ISBN 978-3-211-83686-6 ISBN 978-3-7091-2546-5 (eBook)
DOI 10.1007/978-3-7091-2546-5

PREFACE

Magnetohydrodynamics (MHD) is the study of the interaction between electromagnetic fields and the flow of conducting fluids like liquid metals, molten semiconductors, and plasmas. It presents a fundamental challenge to both fluid dynamics and electrodynamics with applications ranging from materials processing, to the understanding of the origin of Earth's magnetic field and the nature of solar flares.

The present lecture notes are based on lectures given during the tenth IUTAM International Summer School on Magnetohydrodynamics held at CISM in Udine (Italy) from 21 to 25 June 1999. The lecture notes are intended to be accessible to graduate and postgraduate students as well as academics and research engineers who wish to become familiar with magnetohydrodynamics.

The summer school was the brainchild of René Moreau who quietly but efficiently organised the entire event in such a way that it was a delight to lecturers and participants alike. It is a pleasure to acknowledge our gratitude to René Moreau, not just for his extraordinary contribution to this school, but for his unique and sustained contribution to MHD over the last few decades.

We wish also to express our gratitude to the International Centre for Mechanical Sciences (CISM) and the International Union for Theoretical and Applied Mechanics (IUTAM) for their support and the authors for their effort in preparing their lectures in a camera-ready form.

Peter A. Davidson
Andrè Thess

CONTENTS

Chapter I

U. Müller, L. Bühler

LIQUID METAL MAGNETO-HYDRAULICS FLOWS IN DUCTS AND CAVITIES

Liquid Metal Magneto-Hydraulics Flows in Ducts and Cavities

U. Müller and L. Bühler

Forschungszentrum Karlsruhe, Germany

Abstract. This contribution gives an overview on liquid metal flow in engineering applications such as duct flows in various geometries and buoyant flows in cavities. Early results, some of them may be termed classical, are presented as well as results obtained in recent years. It is not the aim to give a complete overview but to introduce the reader to fascinating subject of liquid metal magnetohydrodynamics.

1 Magneto-Hydraulics

1.1 Fundamental Phenomena

Several fundamental phenomena in MHD-duct flow can be explained by simple considerations.

We consider a duct with rectangular cross-section of height $2a$ and width $2d$. The thickness of the duct wall is t_w. An electrically conducting liquid of conductivity σ flows forced by a constant pressure gradient fully developed in the duct with a mean velocity $\mathbf{v}_0$. An external magnetic field $\mathbf{B}_0$ whose direction is parallel to the two channel side walls at lateral distance $2d$ acts on the duct flow.

The magnetic field induces an electric field $\mathbf{E} = \mathbf{v} \times \mathbf{B}$ in the electrically conducting fluid and correspondingly a potential difference between the other two channel side walls, indicated in figure 1 by the symbols $\oplus$ and $\ominus$. The potential difference generates a current in the fluid and in conducting channel walls. The current is generally not uniformly distributed in the fluid and may be locally characterized by the current density $\mathbf{j}$. Currents that flow perpendicular to the magnetic field lines induce a Lorentz-force $\mathbf{F}_L = \mathbf{j} \times \mathbf{B}$, which in our case acts opposite to the flow direction. The current density can be expected to be uniform everywhere in the channel cross section except near the walls. The electrical conductivities of the walls may differ from that of the fluid. Therefore, a nearly uniform Lorentz force acts in the bulk of the flow against the driving external pressure difference. To maintain the volumetric flow rate in the channel the Lorentz forces must be compensated by the driving pressure gradient. MHD channel flows suffer thus from significant electromagnetic pressure losses (in addition to viscous friction losses).

Moreover, the uniform action of the Lorentz force in the core of the channel flattens the parabolically shaped velocity distribution that would be present in the channel center in ordinary friction dominated laminar duct flow. This feature is sketched in figure 1b. For strong external magnetic field the flow turns into a quasi-piston flow. Depending on the electrical conductivity of the channel walls the current density may become non uniform near the walls.

For illustrating the principle effects we consider here three specific cases:

First, the channel wall is perfectly electrically insulating i.e. $\sigma_w = 0$, see figure 2. Because of the condition of charge conservation the currents induced in the bulk flow have to close along the

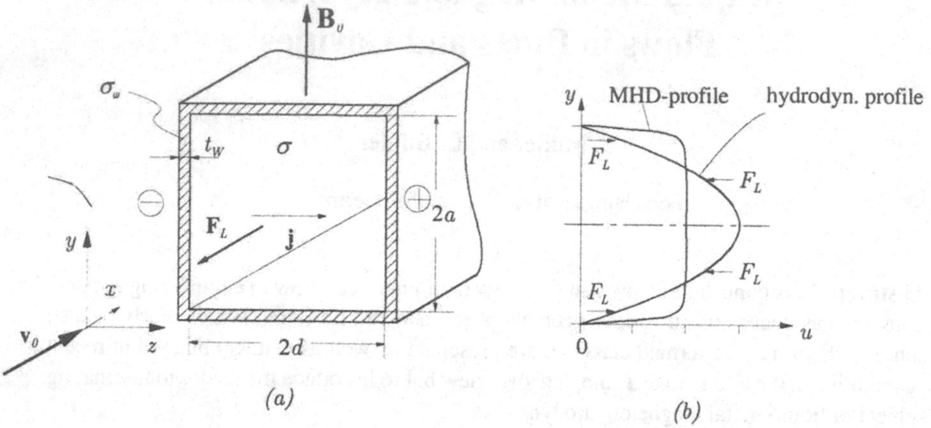

Figure 1. a) Principle sketch of the phenomenology of MHD-channel flow, b) velocity profile at $z = 0$, influence of Lorentz force F_L on the viscous velocity profile.

fluid flow boundary layers near the no slip channel walls. In our specific case the return current flow is nearly parallel with the magnetic field lines at the two side walls and perpendicular to them at the other two channel walls, which are called Hartmann-walls (see figure 2a). In the boundary layer near the Hartmann-walls the Lorentz- and the pressure forces act in the same direction and are balanced by the wall friction forces. This gives rise to very thin velocity boundary layers with steep velocity gradients which match the piston flow profile of the channel core (see figure 2b and 2c). Near the two side walls parallel to the magnetic field the Lorentz forces vanish or are at least very weak because the current flow is (nearly) parallel to the **B**-field. Therefor the pressure forces are balanced by the wall friction forces only and we may expect hydrodynamic behavior where the boundary layer has to match the core flow with the no-slip at the wall. The side wall layers are observed to be thicker than the Hartmann-boundary layer. Their thickness will be quantified later.

Second, the channel wall is electrically perfectly conducting, i.e. $\sigma_w \to \infty$; see figure 3. The currents induced within the fluid close their path through the channel walls and the fluid zones with low velocity near the walls. As the return current encounter no resistance the current density and thus the Lorentz-forces are large in the fluid. A significantly larger electromagnetic pressure loss is to be expected compared to the case of an insulating channel wall. The velocity boundary layers at the Hartmann-walls are very thin as the same force reasoning holds as before. A new phenomenon is observed at the channel walls parallel to **B**, hereafter called side walls. The local velocities in the side wall boundary layer exceed the channel core velocity in a narrow domain located along the side walls (see figure 3b).

Third, the channel walls perpendicular to **B** are perfectly conducting $\sigma_w = \sigma_H = \infty$ while the side walls are insulating, $\sigma_w = \sigma_s = 0$. The short circuit current close in this case mainly through the Hartmann-walls or the velocity boundary layers near the side walls (see figure 4a). In the core the pressure gradient is balanced by the Lorentz forces while near the side walls the pressure gradient is in equilibrium with the friction forces and the much smaller Lorentz

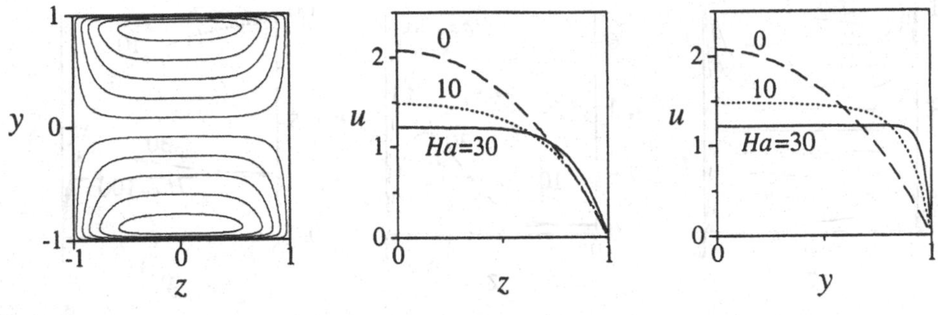

Figure 2. Current paths for MHD-flow in a square channel with electrically insulating walls for $Ha = 30$, velocity distribution along the symmetry axes z and y for different Hartmann numbers.

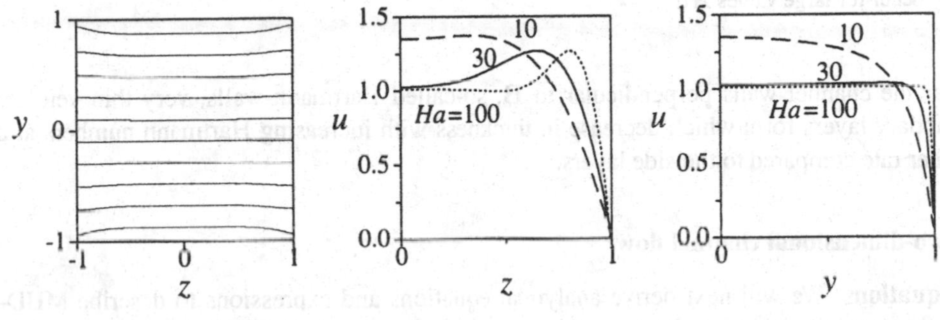

Figure 3. Current paths for MHD-flow in a square channel with electrically well conducting walls for $Ha = 30$, velocity distribution along the symmetry axes z and y for different Hartmann numbers.

forces. Smaller Lorentz forces near the sides allow for considerably higher velocities. A significant amount of the total volume flux may be transported in the so-called side wall jets. The velocity profile along the symmetry axis perpendicular to **B** has a typical shape resembling the letter M and is frequently called M-profile. Accurate theoretical and experimental investigations have shown that in the vicinity of the side layers, towards the channel center, even areas of reverse flow can occur as can be noticed in figure 4b.

The qualitative discussion of the straight channel flow have exposed some very general features of MHD piping networks. For intermediate and strong magnetic fields corresponding to moderate and high Hartmann numbers Ha one finds:

- The current density is nearly constant in the major domain of the channel cross section. This domain may be called the core region. The velocity distribution in this domain is flat.
- Near the side walls parallel to the magnetic field velocity boundary layer form, which may exhibit jet like character. The thickness of the side layer decreases with increasing intensity of **B**, i.e. increasing Hartmann numbers.

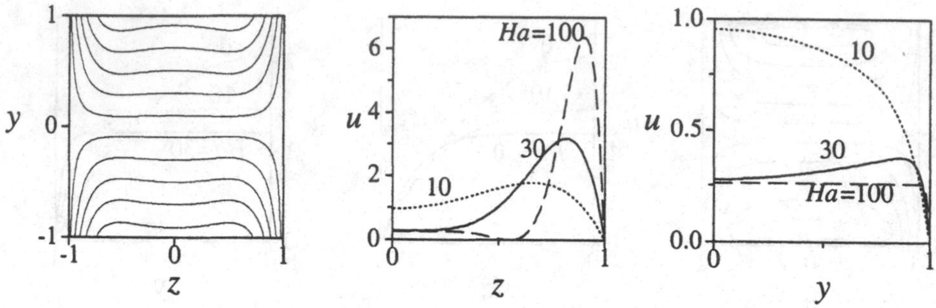

Figure 4. Current paths in square channel with to perfectly conducting Hartmann-walls ($\sigma_H \to \infty$) and two electrically insulating side walls ($\sigma_s = 0$) for $Ha = 30$. Velocity distribution along the symmetry axes z and y for different Hartmann numbers. There are high overspeeds in the side layers and areas of reversed flow may occur for large values Ha.

- Near the channel walls perpendicular to **B**, so-called Hartmann-walls, very thin velocity boundary layers form which decrease in thickness with increasing Hartmann numbers at a higher rate compared to the side layers.

1.2 Two-dimensional channel flow

Basic equations We will next derive analytical equations and expressions to describe MHD-channel flow quantitatively. The basis for this is given by the general equations for incompressible MHD-flow.

We consider a straight channel of arbitrary cross section in a constant homogeneous magnetic field $\mathbf{B} = B_0\hat{\mathbf{y}}$ which is perpendicular to the channel axis. The flow may is driven by some external pressure difference between the beginning and the end of the channel. The arrangement may be seen from figure 1a.

For channels with constant cross section a fully developed flow establishes where the velocity $\mathbf{v} = u\hat{\mathbf{x}}$ has one component only depending on the coordinates y, z. We have

$$\mathbf{v} = [u(y,z), 0, 0]. \tag{1}$$

There is no electric current in the flow direction x and because of Ampère's law, $\mu\mathbf{j} = \nabla \times \mathbf{B}$, a magnetic field component in z-direction is not induced. Furthermore as $\mathbf{B}$ is source free, $\nabla \cdot \mathbf{B} = 0$, only an induced x-component $b(y,z)$ can exist such that

$$\mathbf{B} = [b(y,z), B_o, 0]. \tag{2}$$

Using Ampère's law we obtain

$$j_x = -\frac{1}{\mu}\frac{\partial B_o}{\partial y} \equiv 0, \quad j_y = \frac{1}{\mu}\frac{\partial b}{\partial z}, \quad j_z = -\frac{1}{\mu}\frac{\partial b}{\partial y}. \tag{3}$$

In the equations (3) the induced field b can be interpreted as a stream function representation of the current density $\mathbf{j}$. Therefore isolines of the induced b are streamlines of the steady current density vector field and $\mathbf{j}$ fulfil the charge conservation equation identically in the form

$$\frac{\partial j_y}{\partial y} + \frac{\partial j_z}{\partial z} = 0. \tag{4}$$

It is easily confirmed that the only current component entering the Lorentz force is j_z. Evaluating the momentum equation in the flow direction in particular the Lorentz force $\mathbf{j} \times \mathbf{B}$ we get with equations (3)

$$-\frac{\partial p}{\partial x} + \frac{B_0}{\mu}\frac{\partial b}{\partial y} + \rho\nu\left(\frac{\partial^2 u}{\partial y^2} + \frac{\partial^2 u}{\partial z^2}\right) = 0. \tag{5}$$

where ν is the kinematic viscosity and ρ is the fluid density.

The current is related to the electric field by Ohm's law $j_z = \sigma(E_z + uB_o)$

$$j_y = \sigma E_y, \; j_z = \sigma(E_z + uB_o). \tag{6}$$

The channel flow is now characterized by the equations (4), (6) and (5) with the unknown variables u, b, j_y, j_z, E_z.

We next reduce the number of variables by applying the $\nabla\times$-operator to the current density. This gives

$$\frac{\partial j_z}{\partial y} - \frac{\partial j_y}{\partial z} = -\frac{1}{\mu}\left(\frac{\partial^2 b}{\partial y^2} + \frac{\partial^2 b}{\partial z^2}\right) \tag{7}$$

and furthermore using Ohm's relation

$$\frac{\partial j_z}{\partial y} - \frac{\partial j_y}{\partial z} = \sigma\left[\left(\frac{\partial E_z}{\partial y} - \frac{\partial E_y}{\partial z}\right) + B_o\frac{\partial u}{\partial y}\right]. \tag{8}$$

Because the curl of the electric field vanishes under the condition of time independent magnetic fields the result of equations (7) and (8) is:

$$\mu\sigma B_0\frac{\partial u}{\partial y} + \frac{\partial^2 b}{\partial y^2} + \frac{\partial^2 b}{\partial z^2} = 0. \tag{9}$$

Equations (5) and (9) are the basic equations to determine the distribution of the velocity and the induced magnetic field.

It is useful to transform the equations into a dimensionless form by introduction of scaled variables as follows:

$$(y, z) = a\,(y', z'), \;\; u = v_0 u', \;\; \mathbf{j} = j_0\mathbf{j}', \;\; b = b_0 b', \tag{10}$$

with $v_0 = \frac{a^2}{\rho\nu}\left(-\frac{\partial p}{\partial x}\right)$, $j_0 = \sigma v_0 B_0$, $b_0 = \mu\sqrt{\sigma\rho\nu}v_0$

After dropping the dashes from the dimensionless variables the equations (5) and (9) take the symmetric form

$$Ha\frac{\partial b}{\partial y} + \frac{\partial^2 u}{\partial y^2} + \frac{\partial^2 u}{\partial z^2} = -1, \tag{11}$$

$$Ha\frac{\partial u}{\partial y}+\frac{\partial^2 b}{\partial y^2}+\frac{\partial^2 b}{\partial z^2}=0. \tag{12}$$

where Ha is a dimensionless group called the Hartmann number (Hartmann (1937) investigated first the MHD flow in a plate channel)

$$Ha = aB_0\sqrt{\frac{\sigma}{\rho\nu}}. \tag{13}$$

To evaluate the equations (11) and (12) for the velocity and the induced magnetic field in practical cases adequate boundary conditions have to be provided or derived in particular those for the induced magnetic field.

Hydrodynamic boundary conditions For channel flow we consider viscous liquids with no slip at the channel wall. Therefore all the velocity components vanish at the wall:

$$\mathbf{v}|_{wall} = 0. \tag{14}$$

At the inlet and outlet of the duct a constant pressure is prescribed resulting in a driving pressure difference Δp and a constant pressure gradient

$$\frac{\partial p}{\partial x}=\frac{\Delta p}{\Delta L}=const, \tag{15}$$

with ΔL the duct length.

Electromagnetic boundary conditions The electromagnetic boundary conditions are determined by the electrical conductivity of the channel wall material σ_s. We consider here three typical cases (see also section 1.1).

- electrically insulating walls: $\sigma_w = 0$;
- electrically perfectly conducting walls: $\sigma_w \to \infty$;
- walls of finite, but constant electrical conductivity.

In insulating walls there is no electric current. No current can enter the wall from the liquid side and therefore the normal component of the current density in the liquid vanishes at the wall

$$\mathbf{j}\cdot\mathbf{n}|_{wall} = 0. \tag{16}$$

This condition can be transferred to the induced magnetic field b with the help of Ampére's law and the Stokes' integral formula. We consider a small surface segment A of the channel (inner wall) surface (see figure 5). For the surface segment the following relation holds:

$$0=\int_A \mathbf{j}\cdot\mathbf{n}\,dA=\int_A \frac{1}{\mu}(\nabla\times\mathbf{B})\cdot\mathbf{n}\,dA=\int_C \mathbf{B}\cdot\mathbf{t}\,ds, \tag{17}$$

where C is the contour line of the surface element A and $\mathbf{t}$ is the tangential unit vector to this line. It follows immediately for arbitrarily chosen A that for our case of electrically insulating walls

$$b|_{wall} = 0. \tag{18}$$

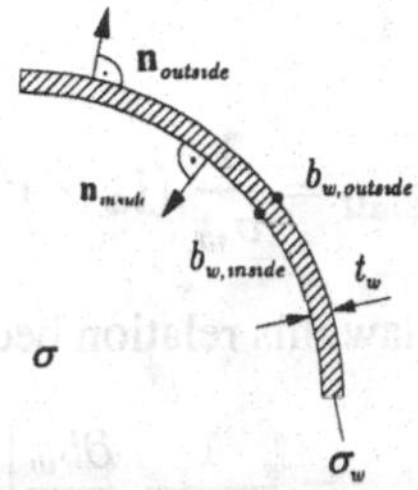

Figure 5. Sketch of a channel wall section, its dimensions and the material properties.

In solids of infinite electrical conductivity the electrical field $\mathbf{E}_w$ vanishes according to Ohm's law, therefore

$$\mathbf{E}|_{wall} = 0. \tag{19}$$

Using Faraday's law in a flat fluid-wall volume element and manipulating it with Stokes' integral formula the requirement is that the tangential components of the electrical field are continuous across the liquid-solid interface (we assume here that no surface charges are located at the interface). This condition can be written as

$$(\mathbf{E} - \mathbf{E}_w) \times \mathbf{n}|_{wall} = 0, \tag{20}$$

where $\mathbf{n}$ is a normal vector to the wall surface.

With equation (19) we get

$$\mathbf{E} \times \mathbf{n}|_{wall} = 0, \tag{21}$$

and using Ohm's law at the wall with $\mathbf{v}|_{wall} = 0$ we obtain

$$(\mathbf{j} \times \mathbf{n})|_{wall} = 0. \tag{22}$$

This relationship means, that the tangential components of the current density vanish at the wall and that the current enters the wall of infinite electrical conductivity in normal direction.

The wall condition for the current density can be transferred to a condition for the induced magnetic field b with the help of Ampère's law. Equation (22) reads then

$$(\nabla \times \mathbf{B}) \times \mathbf{n}|_{wall} = 0. \tag{23}$$

In our specific case the evaluation gives

$$\frac{\partial b}{\partial n} = 0. \tag{24}$$

In technical piping systems channel walls consist frequently of metallic material with finite electrical conductivity σ_w. Moreover, the wall thickness t_w is mostly small compared to the hydraulic diameter a, i.e. $t_w/a << 1$ (see figure 5). In the absence of surface charges at the liquid-solid interface the continuity of the electrical field holds and we have with equation (20) and using Ohm's law

$$\frac{1}{\sigma}\left(\mathbf{j}\times\mathbf{n}\right)\Big|_{wall} = \frac{1}{\sigma_w}\left(\mathbf{j}_w\times\mathbf{n}\right)\Big|_{wall}. \tag{25}$$

Furthermore, with the help of Ampère's law this relation becomes

$$\frac{1}{\mu\sigma}\left.\frac{\partial b}{\partial n}\right|_{wall} = \frac{1}{\mu_w\sigma_w}\left.\frac{\partial b_w}{\partial n}\right|_{wall} \tag{26}$$

This means that the normal derivatives of the components of the induced magnetic field are continuous.

In case of thin channel walls the induced magnetic field in the wall can be expanded as a Taylor series from which we can approximate the gradient

$$\frac{\partial b_w}{\partial n} = \frac{1}{t_w}\left(b_{outside} - b_{inside}\right) \tag{27}$$

where $b_{outside}$ and b_{inside} are the values at the outside and inside surface of the channel wall (see figure 5). As there are no currents in the insulating domain outside the channel wall the induced field vanishes, $b_{outside} = 0$. Using equations (26) and (27) the boundary condition for the induced magnetic field can be written as

$$\frac{\partial b}{\partial n} + \frac{\mu\sigma}{\mu_w\sigma_w}\frac{1}{t_w}b = 0 \tag{28}$$

This relationship can be transformed in a dimensionless form by scaling the coordinates with the channel dimension a. We than arrive at the final form of the boundary condition:

$$\frac{\partial b}{\partial n} + \frac{1}{c}b = 0 \tag{29}$$

with the wall conductance parameter

$$c = \frac{\sigma_w t_w}{\sigma a}. \tag{30}$$

This parameter describes the ratio of the electrical resistances of the wall and fluid material. As outlined schematically in section 1.1 this parameter will strongly influence the hydromagnetic pressure losses in duct flows because of the effect of the Lorentz forces.

1.3 Basic equations for duct flow using the velocity and the electrical potential as governing variables

For calculating MHD-duct flow in piping systems with conducting duct walls it is often useful to work in the variables velocity $\mathbf{v}$ and electrical potential ϕ instead of using the velocity together with the induced magnetic field, in particular if numerical methods have to be employed to obtain solutions in complex duct geometries such as bends or branching conduits. If the induced magnetic field is negligibly small compared to the external magnetic field, that is the magnetic Reynolds number is very small, the basic MHD-equations read as dimensionless conservation equations for momentum, mass, charge:

$$\frac{1}{N}\left[\partial_t + (\mathbf{v}\cdot\nabla)\right]\mathbf{v} = -\nabla p + \frac{1}{Ha^2}\nabla^2\mathbf{v} + \mathbf{j}\times\mathbf{B}, \tag{31}$$

$$\nabla\cdot\mathbf{v}{=}0, \tag{32}$$

$$\nabla\cdot\mathbf{j}{=}0. \tag{33}$$

Furthermore Ohm's law holds

$$\mathbf{j} = -\nabla\phi + \mathbf{v}\times\mathbf{B}, \tag{34}$$

where the electric field is expressed by the gradient of the scalar electric potential ϕ as $E = -\nabla\phi$. The dimensionless groups are as previously introduced the interaction parameter N and the Hartmann number Ha. In the case of a uniform external magnetic field, $\mathbf{B} = \hat{\mathbf{y}}$ is a unit vector in the direction of the magnetic field, chosen e.g. as the y-direction. For engineering applications the scale for velocity v_0 is often used as the average velocity at some cross section, in contrast to the scale introduced earlier in a previous subsection (compare equation 10). Using the charge conservation equation (33) equation (34) can be transferred to the relation

$$\nabla^2\phi = \nabla\cdot(\mathbf{v}\times\mathbf{B})\,. \tag{35}$$

Inserting equation (34) into equation (31) and expanding the cross products gives

$$\frac{1}{N}\left[\partial_t + (\mathbf{v}\cdot\nabla)\right]\mathbf{v} = -\nabla p + \frac{1}{Ha^2}\nabla^2\mathbf{v} - \mathbf{v}_\perp + \mathbf{B}\times\nabla\phi \tag{36}$$

where the vector field $\mathbf{v}_\perp$ represents the two velocity components perpendicular to the direction of the external magnetic field.

This set of equations (31)-(33) and (35) has to be solved for adequate kinematic and electrical boundary conditions. The kinematic boundary condition is $\mathbf{v}|_w = 0$. For insulating surfaces we have that

$$\mathbf{j}\cdot\mathbf{n}|_w = 0 \tag{37}$$

which gives a vanishing normal derivative of the potential

$$\left.\frac{\partial\phi}{\partial n}\right|_w = 0. \tag{38}$$

For a perfectly conducting wall the potential at the wall becomes uniform since potential differences cannot exist. The value of wall potential then can be put to zero without loosing any information. For finite conducting thin walls (known wall and fluid conductivity) the local current entering the wall is discharged into the thin wall in a quasi two dimensional way. This is described by the relation

$$\frac{\partial j_n}{\partial n} = -\nabla_\perp \cdot \mathbf{j}, \tag{39}$$

where the gradient vector $\nabla_\perp$ is in the tangential plane to the wall. With Ohm's law in the wall, integrated in the wall normal direction while knowing that the potential does not vary across the wall to the leading order of approximation one finds finally the relationship

$$\mathbf{j} \cdot \mathbf{n}|_w = -\frac{\partial \phi}{\partial n} = \nabla \cdot (c \nabla \phi_w) \tag{40}$$

where c is the wall conductance parameter.

For two dimensional fully developed channel flow the boundary value problem is therefore adequately described by the following set of equations and boundary conditions

$$-\frac{dp}{dz} + \frac{1}{Ha^2}\left(\frac{\partial^2 u}{\partial y^2} + \frac{\partial u}{\partial z^2}\right) - u + \frac{\partial \phi}{\partial z} = 0, \tag{41}$$

$$\frac{\partial^2 \phi}{\partial y^2} + \frac{\partial^2 \phi}{\partial z^2} = \frac{\partial u}{\partial z}, \tag{42}$$

$$u = 0, \quad \frac{\partial \phi}{\partial n} = -\nabla \cdot (c \nabla \phi_w) \quad \text{at the wall.} \tag{43}$$

This set is frequently used either for full numerical simulation of channel flow but it may also serve as basis for asymptotic analysis approximation.

2 Analytical solutions for MHD-channel flow

There are two classes of analytical solutions for two-dimensional MHD-channel flow. The first class, limited in number by certain channel geometries, comprises the exact solutions which may be represented by series expansion of eigenfunctions. A second class of solutions may be obtained as an asymptotic approximation for large Hartmann number $Ha \gg 1$. Solution of both classes are derived and discussed during the next sections.

2.1 Flow in a plate channel – Hartmann flow

Hartmann (1937) first investigated experimentally and theoretically the MHD-flow in the gap between two parallel plates (see figure 6). This investigation provided fundamental knowledge for the development of several MHD-devices such as MHD-pumps, generators, brakes and flow meters. Figure 6 shows the geometrical arrangement. The external magnetic field is perpendicular to the two channel walls. The lateral channel walls are at infinity.

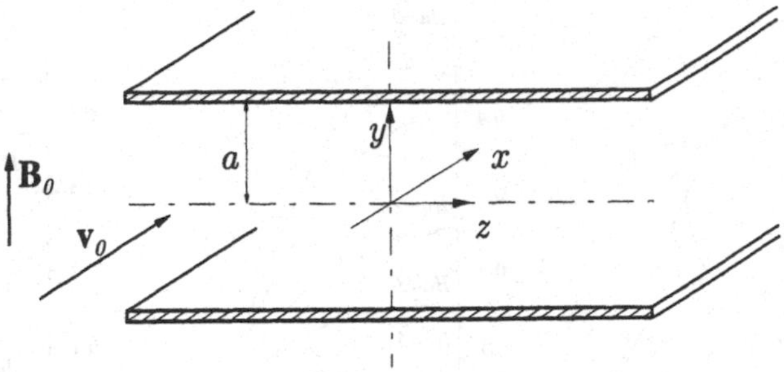

Figure 6. Sketch of a plate channel.

Therefore the relevant dependent variables, the velocity u and induced magnetic field b depend only on one independent variable y. The problem generally described by the set of equations (11), (12) and boundary conditions (14) and (29) reduces to the simpler form

$$Ha\frac{\partial b}{\partial y} + \frac{\partial^2 u}{\partial y^2} = -1,$$

$$Ha\frac{\partial u}{\partial y} + \frac{\partial^2 b}{\partial y^2} = 0, \quad \text{for} -1 < y < 1 \tag{44}$$

and

$$u = 0, \qquad \frac{\partial b}{\partial y} + \frac{1}{c}b = 0 \quad \text{at } y = \pm 1; \tag{45}$$

Here $c \to 0$ corresponds to electrically insulating walls and $c \to \infty$ to electrically perfectly conducting walls.

Solutions of the general form

$$u\,(y) = A\,[\cosh\,(Ha\;y) - \cosh\,(Ha)] \tag{46}$$
$$b\,(y) = B_1 \sinh\,(Ha\;y) + B_2 y \tag{47}$$

are easily confirmed. From the boundary conditions for b at $y = \pm\,1$ we obtain the values for the integration constants B_1 and B_2. Finally by inserting the solutions for u and b in equation (44b) we fix A and get

$$u\,(y) = \hat{u}\left[1 - \frac{\cosh\,(Ha\;y)}{\cosh\,(Ha)}\right],$$

$$b\,(y) = -\frac{y}{Ha} + \hat{u}\frac{\sinh\,(Ha\;y)}{\cosh\,(Ha)}, \tag{48}$$

with the characteristic magnitude of velocity

$$\hat{u} = \frac{1}{Ha}\frac{c+1}{cHa + \tanh\,(Ha)}. \tag{49}$$

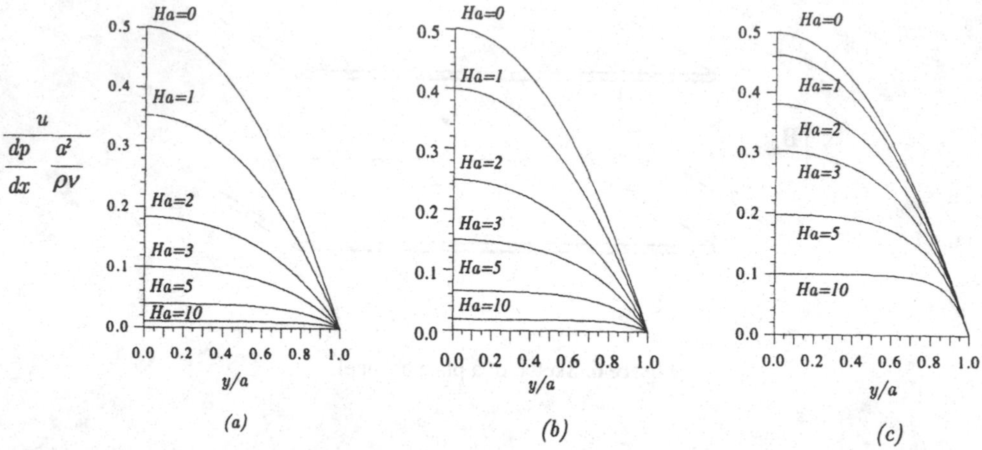

Figure 7. Velocity profiles in dimensionless form for a Hartmann flow. The profiles are presented for Hartmann numbers $Ha = 0, 1, 2, 3, 5, 10$ and for a) electrically perfectly conducting walls ($c \to \infty$), b) for wall of finite conductivity with $c = 1$, c) insulating walls ($c = 0$).

This form of the solution is due to Chang and Lundgren (1961). The velocity profile is shown in figure 7 for different values of the Hartmann number and for three specific selections of c namely$c = 0$, insulating walls; $c = \infty$, perfectly conducting walls; $c = 1$ a finite electrical wall conductivity.

The following characteristic features can be identified.

- For a fixed pressure gradient and fixed values of wall properties the velocity decreases with increasing Hartmann numbers i.e. increasing strength of the external magnetic field.
- With increasing Hartmann numbers (magnetic field strength) the velocity profile flattens in the channel core.
- For a fixed pressure gradient and a given Hartmann number (magnetic field strength) the volumetric flow rate is largest in a channel with electrically insulating walls and smallest with perfectly conducting walls.

We next discuss briefly the two limiting cases of a vanishing external magnetic field, $B_0 = 0$, i.e. $Ha = 0$ and the case for extremely strong external magnetic field, $Ha \gg 1$.

For $Ha \to 0$ we expand the hyperbolic functions in the solution for u in a power series

$$\cosh(Ha\, y) = 1 + \frac{1}{2}(Ha\, y)^2 + \frac{1}{24}(Ha\, y)^4 + O\left[(Ha\, y)^6\right], \tag{50}$$

$$\tanh(Ha) = Ha - \frac{1}{3}Ha^3 + \frac{2}{15}Ha^5 + O\left[(Ha\, y)^7\right], \tag{51}$$

and find in the limiting case, when $Ha \to 0$ the solution for hydrodynamic Poiseuille flow,

$$\lim_{Ha \to 0} u(y) = \frac{1}{2}\left(1 - y^2\right). \tag{52}$$

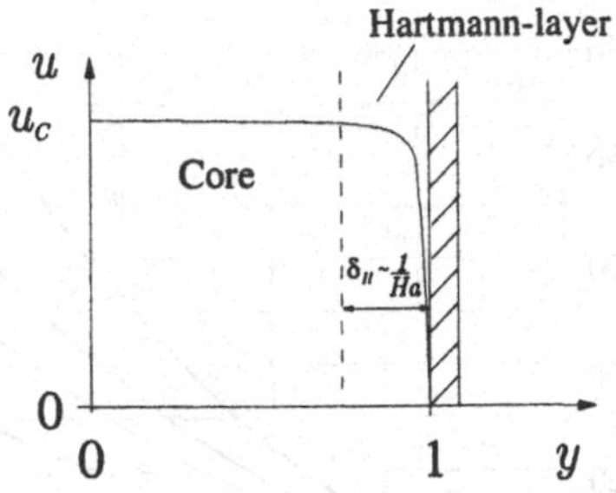

Figure 8. Sketch of a Hartmann velvety profile for high Hartmann numbers $Ha >> 1$, the thickness of the Hartmann boundary layer scales as $\delta \sim Ha^{-1}$.

For $Ha \gg 1$ the hyperbolic functions asymptote to

$$\cosh(Ha\,y) \to \frac{1}{2}\exp(Ha\,y), \quad \cosh(Ha) \to \frac{1}{2}\exp(Ha), \quad \tanh(Ha) \to 1. \tag{53}$$

Introducing these expressions into equations (48a,b) we get

$$u(y) = u_c\{1 - \exp[Ha(|y| - 1)]\},$$
$$b(y) = -\frac{y}{Ha} \pm u_c \exp[Ha(|y| - 1)] \quad \text{for } y \gtrless 0, \tag{54}$$

with

$$u_c \to \frac{1}{Ha}\frac{c+1}{cHa+1}. \tag{55}$$

These relations show for $Ha \gg 1$ an exponential decrease of the velocity and the induced magnetic field in the vicinity of the channel walls. This gives rise to thin boundary layers, whose thickness δ_H is of the order

$$\delta_H = O\left(Ha^{-1}\right). \tag{56}$$

The boundary layers of exponential character (see figure 8) are called Hartmann layers. They can be generally found at all channel walls where the magnetic field has a non zero component normal to the wall.

From the equations (54) we also recognize that the velocity and the induced magnetic field behave in the core of the channel like

$$u_c = \frac{1}{Ha}\frac{c+1}{cHa+1}, \tag{57}$$

$$b_c = -\frac{1}{Ha}y. \tag{58}$$

These expressions are frequently referred to as "core" or "bulk" velocity and magnetic field respectively (see figure 8). We obtain from the relations also the special cases for insulating and perfectly conducting channel walls with $c = 0$ and $c \to \infty$. Then the core velocity is

$$\begin{aligned} u_c &= Ha^{-1} \\ u_c &= Ha^{-2} \end{aligned} \quad \text{for} \quad \begin{aligned} &c = 0, \\ &c \to \infty. \end{aligned} \tag{59}$$

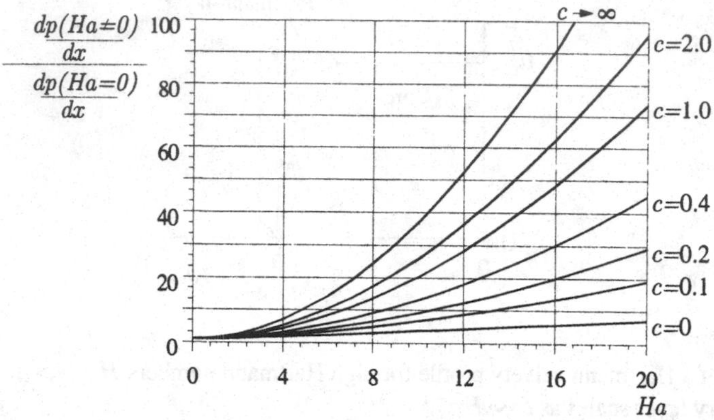

Figure 9. Dimensionless MHD-pressure losses as a function of the Hartmann number Ha and the wall conductance parameter c.

Volumetric flow rate and pressure losses Knowing the velocity distribution across the plate channel the volumetric flow rate can be calculated by an integration. We get the dimensionless flow rate

$$Q = \int_{-1}^{1} u\,(y)\,dy = 2\frac{c+1}{Ha^2}\frac{Ha - \tanh\,(Ha)}{cHa + \tanh\,(Ha)} \tag{60}$$

Because, of the velocity scaling according to equation (10) the nondimensional Q is independent of the scaled pressure gradient, since the latter one evaluates to unity.

For discussing further the hydromagnetic pressure losses for a fixed flow rate compared to the hydraulic case we form the ratio $Q(Ha = 0)/Q$ and use equation (60). We get

$$\frac{Q(Ha \neq 0)}{Q(Ha = 0)} = \frac{\frac{\partial p}{\partial x}\,(Ha = 0)}{\frac{\partial p}{\partial x}\,(Ha \neq 0)} = \frac{c+1}{Ha^2}\frac{Ha - \tanh\,(Ha)}{cHa + \tanh\,(Ha)}. \tag{61}$$

The following obvious conclusions hold:

- With increasing Hartmann number an increasing pressure gradient is required to keep the flow rate constant.
- For maintaining a fixed flow rate a higher pressure gradient is required in a channel with well conducting walls than in one with poorly conducting walls for a fixed Hartmann number.

Figure 9 demonstrates the various dependencies of the dimensionless pressure gradient on the Hartmann number and the wall conductance parameter. It is evident from this graph that for perfectly conducting walls ($c \to \infty$) the dimensionless pressure losses vary quadratically with Ha while this dependency is only linear for insulating walls ($c = 0$).

Comparison Theory and Experiment A plate channel geometry can hardly by realized in an experiment but a reasonable approximation to the ideal case can be obtained by employing a flat

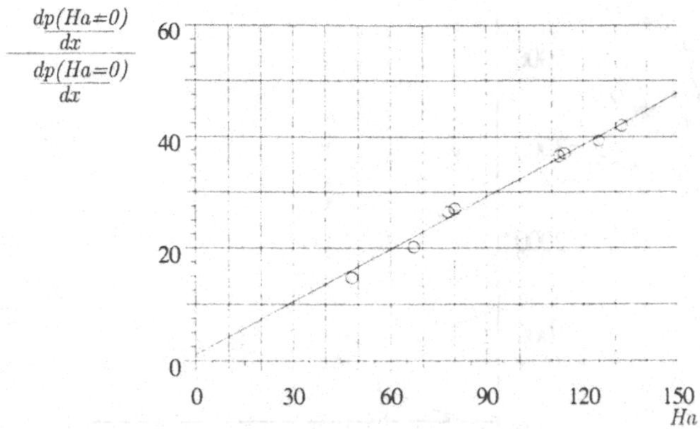

Figure 10. Experimental results for pressure drop (○) in a flat channel with aspect ratio $d/a = 15$ compared with the analytical solution for Hartmann flow (Murgatroyd (1953)).

rectangular duct with a large aspect ratio width versus height d/a. The aspect ratio should be of the order $d/a \gtrsim 10$.

Local velocity measurements are generally difficult to perform since velocity measurements in liquid metals, which are commonly used as test fluids in MHD-investigations, require intrusive instrumentation, e.g. Pitot tubes or permanent magnet probes, which may significantly perturb the flow itself. Therefore, commonly differences of the static pressure between inlet and outlet of the channel are measured. Murgatroyd (1953) investigated the MHD-flow in an electrically insulated rectangular channel of aspect ratio $d/a = 15$ for Hartmann numbers $40 < Ha < 130$. He found a good agreement of the measured pressure losses with the theoretical predictions of the Hartmann flow. The comparison is displayed in figure 10.

To realize an experiment using a channel with perfectly conducting walls is even more difficult from a practical point of view, since the electrical conductivities of metals are all of the same order of magnitude. A reasonable approximation to the case of perfectly conducting channel walls was achieved Branover et al. (1967) who employed mercury as the test fluid and copper as the wall material. The experiment was conducted in a rectangular channel of aspect ratio $d/a = 14$. The comparison between the measurements and the theoretical predictions is shown in figure 11. The agreement between the measured data points and the theoretical values are particularly good for small Hartmann numbers. A certain deviation at larger values Ha can be attributed to the effect of the confining side walls which do influence the pressure losses by the formation of overspeeds in their vicinity (see section 1.1).

The Hartmann flow as electric generator or pump So far we considered the Hartmann flow for a given fixed external pressure gradient. This pressure gradient drives the flow and induces an electric field perpendicular to the flow and the magnetic field direction. This gives rise to potential differences and electric current flows along a transverse direction (here the z-direction), parallel to the channel walls. The flow acts as electrical power generator.

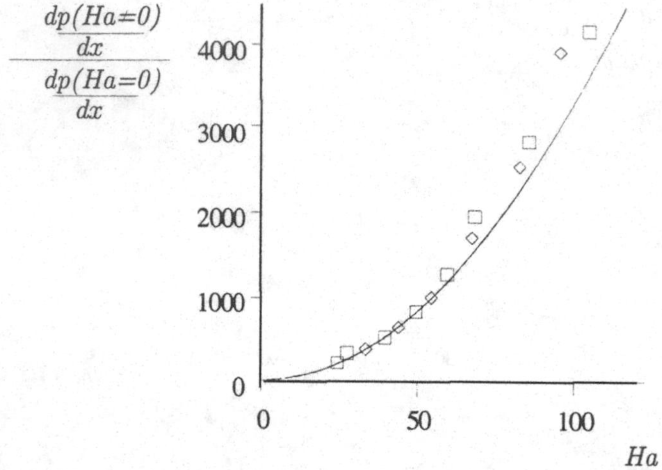

Figure 11. Experimental results for pressure drop in a flat channel with aspect ratio $d/a = 14$ with high wall conductivity $c = 100$ compared with the analytical solution for Hartmann flow. (◇) $Re = 1480$, (□) $Re = 2980$ (Branover et al. (1967)).

On the other hand an externally supplied voltage along the same transverse axis resulting in forced electric current will generate a Lorentz force density in the channel cross section which will generate a pressure difference and drive or brake a fluid flow. We shall next discuss the set of parameters for which the channel acts as a generator or a pump. The situation is sketched in figure 12.

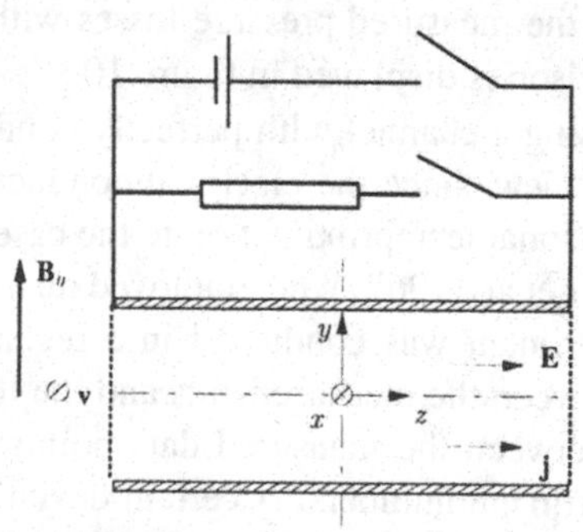

Figure 12. Electric circuit diagram for MHD-channel flow

In the fluid an integration along y of Ohm's law

$$j_z = E + u. \tag{62}$$

leads to

$$I = E + Q, \tag{63}$$

with the flow rate Q and the z-component of the electric field E, from which the subscript is dropped here for simplicity. The net current I stands for the integrated current density in the fluid and in the wall. This equation can be evaluated after the velocity $u(y)$ is known. In the previous subsection a solution has been derived that assumed that all current returns through the Hartmann layers and through the walls. In that case of pure Hartmann flow I was absent, $I = 0$. If a net current is allowed now the solution will deviate from the latter one. With a net current $I \neq 0$ the Shercliff thin wall condition modifies to give now

$$\frac{\partial b}{\partial y} + \frac{1}{c}\left(b + \frac{1}{2}HaI\right) = 0. \tag{64}$$

The solution of the basic equations now leads to new integration constants and the flow rate finally becomes

$$Q = 2\frac{1}{Ha^2}\frac{Ha - \tanh(Ha)}{cHa + \tanh(Ha)}\left(1 + c - \frac{1}{2}Ha^2 I\right). \tag{65}$$

One can identify the part of flow rate that is independent of I. It is present even if there exists no external circuit, when all switches are open, when $I = 0$. This part of flow rate is called the Hartmann flow rate Q_H which can be used for a convenient representation of the result as

$$\frac{Q}{Q_H} = 1 - \frac{1}{2}Ha^2 I\frac{1}{1+c}. \tag{66}$$

Inserting this result into equation (63) one finds especially for large Hartmann numbers, as $Q_H \to \frac{2}{Ha^2}\frac{1+c}{c}$ for $Ha \to \infty$, the result

$$\frac{E}{Q_H} = -1 + \frac{1}{2}Ha^2 I. \tag{67}$$

Knowing these results one can draw a picture of the situation as shown in figure 13. If the circuit is open, when $I = 0$, one recovers the Hartmann solution with $Q/Q_H = 1$, $E/Q_H = -1$. This case is important for applications of MHD as flow meters, since the electric field which can be measured easily is directly proportional to the flow rate.

The duct flow creates an electric field that may be used as a voltage supply or electrical generator to support external circuits with electrical power. One can extract from the system electrical energy as long as the product EI is negative. The range for which an electrical generator exists is therefore restricted to $0 < \frac{1}{2}Ha^2 I < 1$.

Currents larger than $\frac{1}{2}Ha^2 I = 1$ are possible only if electrical power is supplied to the system from some external source. In these cases the flow is suppressed strongly and the system acts like a brake. The externally supplied electrical power now drives the flow in the direction of the applied pressure gradient and acts like a pump. If $\frac{1}{2}Ha^2 I > 1 + c$ the flow becomes even reversed.

The system behaves like a pump also for the case when the current becomes negative, when $I < 0$. In this case the externally supplied electrical power drives the flow in the same direction as the applied pressure gradient and increases the flow rate compared to the Hartmann flow.

Note, it is not possible to convert all mechanical power into electrical power or vice versa. Especially at large currents a significant fraction of power will be dissipated by Joule heating in the fluid and in the walls.

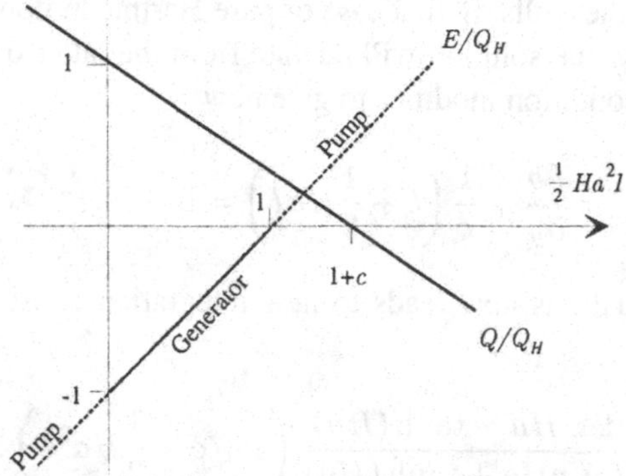

Figure 13. Sketch of electric field and flow rate depending on an externally supplied or extracted current. For $EI < 0$ electric power is extracted and the flow acts like an electrical generator. In the other regions the flow is pumped by input of electric power.

2.2 Flow in rectangular channels

Even for simple rectangular channel geometries there are very few wall-fluid flow arrangements for which analytical solutions have been obtained. Solutions are known for electrically perfectly conducting and perfectly insulating channel walls. No analytical solution in closed form has been obtained yet for the case of a finite wall conductivity. Such a situation or generally inhomogeneous wall conductivities have to be solved by numerical methods e.g. finite difference methods or by asymptotic analysis.

The general discussion in chapter 1.1 show that the behavior of a MHD-channel flow is dominated by the distribution of the current density in the channel cross section and in the confining walls. Special cases for which analytical solutions in form of eigenfunction expansions have been given are shown in figure 14.

Insulating walls We will next derive an analytical solution for an MHD-flow in a rectangular channel with insulating walls. The geometrical dimensions and the MHD-conditions are sketched

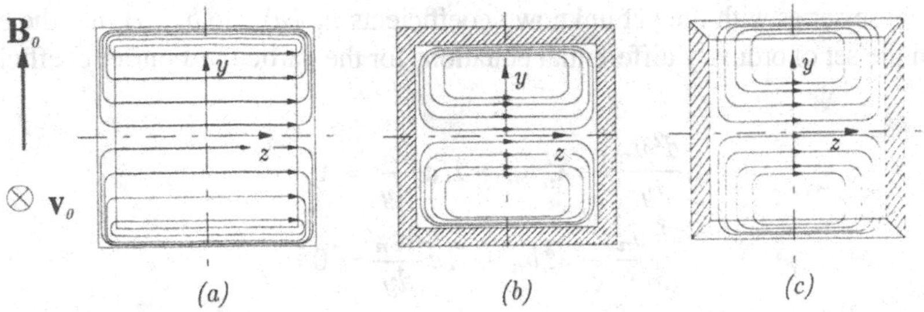

Figure 14. Schematic drawing of the current distribution in a square channel with three different choices of wall conductivities σ_w. a) perfectly conducting walls, b) perfectly insulating walls, c) Hartmann walls perfectly conducting, side wall perfectly insulating. Solutions have been given by a) Ufland (1961) , b) Shercliff (1953), c) Hunt (1965).

in figure 14b. The problem is described by equations which were derived in section 1.2.

$$Ha\frac{\partial b}{\partial y} + \frac{\partial^2 u}{\partial y^2} + \frac{\partial^2 u}{\partial z^2} = -1,$$
$$Ha\frac{\partial u}{\partial y} + \frac{\partial^2 b}{\partial y^2} + \frac{\partial^2 b}{\partial z^2} = 0, \tag{68}$$

with the no-slip boundary condition

$$u\,(y = \pm 1, z) = u\,(y, z = \pm d) = 0 \tag{69}$$

at insulating walls

$$b\,(y = \pm 1, z) = b\,(y, z = \pm d) = 0. \tag{70}$$

In the following for simplicity the analysis is restricted to square ducts with $d = 1$. An extension to rectangular ducts is straight forward.

From symmetry considerations and by taking into account the imposed boundary conditions the solution may be expanded in Fourier series

$$u\,(y, z) = \frac{1}{2}\left[1 - z^2\right] + \sum_{n=0}^{\infty} u_n\,(y) \cos\left(\lambda_n z\right),$$
$$b\,(y, z) = \sum_{n=0}^{\infty} b_n\,(y) \cos\left(\lambda_n z\right), \tag{71}$$

with

$$\lambda_n = (2n + 1)\,\frac{\pi}{2}. \tag{72}$$

Inserting these series with the yet unknown coefficients $u_n(y)$ and $b_n(y)$ into the equations (68) results in the set of ordinary differential equations for the particular Fourier coefficients. We obtain

$$\frac{d^2u_n}{dy^2} - \lambda_n^2 u_n + Ha\frac{db_n}{dy} = 0,$$
$$\frac{d^2b_n}{dy^2} - \lambda_n^2 b_n + Ha\frac{du_n}{dy} = 0. \tag{73}$$

From these equations either u_n or b_n can be eliminated to give an ordinary differential equation of fourth order for either of them, for example for u_n in the form

$$\frac{d^4u_n}{dy^4} - \left(2\lambda_n^2 + Ha^2\right)\frac{d^2u}{dy^2} + \lambda_n^4 u_n = 0. \tag{74}$$

This equation has general solutions of exponential form $u_n = \exp(py)$,where p is a root of the following algebraic equation

$$p_n^4 - \left(2\lambda_n^2 + Ha^2\right)p_n^2 + \lambda_n^4 = 0. \tag{75}$$

Generally there are four solutions to this equation and, therefore, there are four independent solutions to the equations (74). Since the solution for the velocity must be symmetric with regard to the z-coordinate axis the ansatz for u_n can be chosen as

$$u_n(y) = A_n \cosh(p_{n1}y) + B_n \cosh(p_{n2}y) \tag{76}$$

with

$$\left(p_n^2\right)_{1,2} = \lambda_n^2 + \frac{Ha^2}{2} \pm Ha\left(\lambda_n^2 + \frac{Ha^2}{4}\right)^{1/2}. \tag{77}$$

An equivalent solution for the coefficient $b_n(y)$ is obtained from equations (73a) by inserting ansatz (76) and integrating for y. We obtain

$$b_n(y) = \frac{1}{Ha}\left[A_n\frac{\lambda_n^2 - p_{n1}^2}{p_{n1}}\sinh(p_{n1}y) + B_n\frac{\lambda_n^2 - p_{n2}^2}{p_{n2}}\sinh(p_{n2}y)\right] \tag{78}$$

The yet unknown constants of integration A_n and B_n have to be determined next with the help of the boundary conditions for u and b at the wall position $y = \pm 1$.

As A_n and B_n represent coefficients in a Fourier expansion of our solution it is useful to expand the hydrodynamic constituent of the solution-ansatz (71a) into a Fourier series, that is in $-1 \leq z \leq 1$

$$\frac{1}{2}\left(1 - z^2\right) = \sum_{n=0}^{\infty} u_{n0}\cos(\lambda_n z). \tag{79}$$

An evaluation of the coefficients u_{n0} gives

$$u_{n0} = (-1)^n \frac{2}{\lambda_n^3}. \tag{80}$$

The general solution for the velocity can be written as

$$u(y,z)=\sum_{n=0}^{\infty}\left[u_{n0}+A_n\cosh(p_{n1}y)+B_n\cosh(p_{n2}y)\right]\cos(\lambda_n z) \tag{81}$$

The adjustment to the boundary conditions $u(y=\pm 1, z)=0$ gives

$$u_{n0}+A_n\cosh(p_{n1})+B_n\cosh(p_{n2})=0. \tag{82}$$

A corresponding adaption of $b(y)$ to the boundary condition $b(y\pm 1,z)=0$ supplies the relation

$$A_n p_{n2}\left(\lambda_n^2-p_{n1}^2\right)\sinh(p_{n1})+B_n p_{n1}\left(\lambda_n^2-p_{n2}^2\right)\sinh(p_{n2})=0 \tag{83}$$

From equations (82) and (83) the following expressions are obtained for the yet unknown coefficients A_n and B_n:

$$A_n=\frac{p_{n1}\left(\lambda_n^2-p_{n2}^2\right)\sinh(p_{n2})}{\Theta_n}u_{0n}, \tag{84}$$

$$B_n=-\frac{p_{n2}\left(\lambda_n^2-p_{n1}^2\right)\sinh(p_{n1})}{\Theta_n}u_{0n}, \tag{85}$$

where

$$\Theta_n=p_{n2}\left(\lambda_n^2-p_{n1}^2\right)\sinh(p_{n1})\cosh(p_{n2})-p_{n1}\left(\lambda_n^2-p_{n2}^2\right)\sinh(p_{n2})\cosh(p_{n1}). \tag{86}$$

The final analytical form of the solution reads as

$$\begin{aligned} u(y,z)&=\frac{1}{2}\left[1-z^2\right]+2\sum_{n=0}^{\infty}\frac{(-1)^n}{\lambda_n^3\Theta_n}\alpha_n(y)\cos(\lambda_n z),\\ b(y,z)&=2Ha\sum_{n=0}^{\infty}\frac{(-1)^n}{\lambda_n\Theta_n}\beta_n(y)\cos(\lambda_n z), \end{aligned} \tag{87}$$

where

$$\alpha_n(y)=p_{n1}\left(\lambda_n^2-p_{n2}^2\right)\sinh(p_{n2})\cosh(p_{n1}y)-p_{n2}\left(\lambda_n^2-p_{n1}^2\right)\sinh(p_{n1})\cosh(p_{n2}y), \tag{88}$$

and

$$\beta_n(y)=-\sinh(p_{n2})\sinh(p_{n1}y)+\sinh(p_{n1})\sinh(p_{n2}y). \tag{89}$$

Solutions for velocity for different Hartmann numbers $Ha=0,10,30$ are shown already in figure 2. The velocity distribution becomes flat in the channel core with increasing Hartmann numbers, while simultaneously boundary layers with steep velocity gradients form. It is also evident from the pictures, that the Hartmann-boundary layers are significantly thinner compared to the boundary layers at the walls parallel to the external magnetic field, the so called side-layers.

Figure 2 show the isolines of the induced magnetic field b which also represent the "stream lines" of the current density. The figure clearly demonstrates the formation of current boundary layers at the channel walls which resemble even "current sheets" especially at the Hartmann walls.

The volumetric flow rate in the channel is evaluated according to the relation

$$Q = 4 \int_0^1 \int_0^1 u(y,z)\, dydz \tag{90}$$

and results in

$$Q = \frac{4}{3} + 16Ha \sum_{n=0}^{\infty} \frac{\left(\lambda_n^2 + \frac{1}{4}Ha^2\right)^{1/2}}{\lambda_n^4 \Theta_n} \sinh(p_{n1}) \sinh(p_{n2}) \,. \tag{91}$$

The explicit evaluation of this solution turns out to be tedious for large Hartmann numbers as the hyperbolic functions become very large and as the coefficients of the Fourier expansion have alternating signs.

Local velocity measurements are difficult to perform in channels of relatively small cross sections in particular for liquid metal flow. But pressure loss and flow rate measurements can be made with reasonable effort. Such measurements are available. It is common use to represent the pressure losses in terms of a dimensionless loss coefficient

$$\lambda = \frac{dp/dx}{\frac{1}{2}\rho v_0^2} . \tag{92}$$

For large Hartmann numbers $Ha \gg 1$ the expression for the electromagnetic pressure loss coefficient can be evaluated asymptotically as shown by Hunt and Stewartson (1965). Without giving the details one arrives at the following expression

$$\lambda = 2\frac{Ha}{Re}\left(1 - \alpha\frac{a}{d}Ha^{-1/2} - Ha^{-1}\right)^{-1}, \quad \text{for } Ha \gg 1. \tag{93}$$

The coefficient α evaluates to

$$\alpha = \frac{1}{2}\sqrt{2}\frac{\left(-\frac{1}{4}\right)!}{\left(\frac{1}{4}\right)!} \approx 0.95598. \tag{94}$$

One can identify the three contributions to the pressure drop in the brackets of equation (93). The first term is caused by the core flow, the second by the side layers and the last one by the Hartmann layers. To elaborate the dependency of the pressure losses on the magnetic field intensity B_0, say the Hartmann number, it is useful to introduce a ratio of loss coefficients λ/λ_0 where λ_0 is the loss coefficient for the steady laminar viscous channel flow. The viscous laminar channel flow is used as a reference state. Even for higher Reynolds numbers $Re >> 2000$ the MHD flow stays laminar under the influence of strong magnetic fields. Figure 15a shows the dependency of this loss factor ratio as a function of the Hartmann number for rectangular channels of different aspect ratios and insulating walls in a comparison between measured data and theoretical prediction.

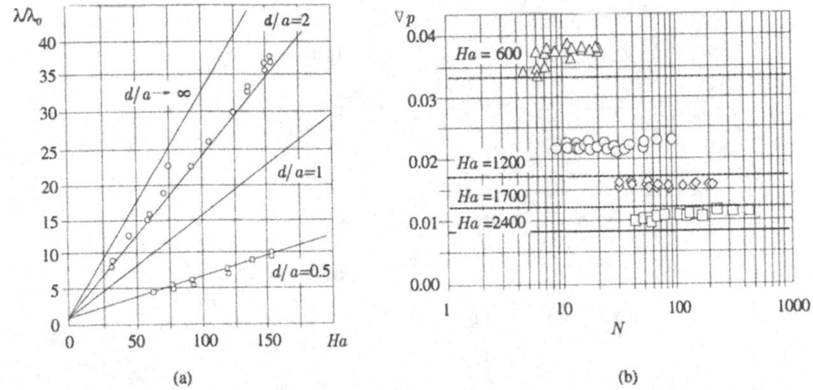

Figure 15. Comparison between measured and calculated pressure losses in a rectangular channel with insulating walls. a) loss coefficient ratio as a function of the Hartmann number Ha for different aspect ratios; measurements: (0) Murgatroyd (1953) for $d/a = 15.5$, (o) Branover (1978), (□) ? (?), calculations: (-). b) dimensionless electromagnetic pressure gradient for different Hartmann number as a function of the interaction parameter $N = a\sigma B_0^2/(\rho v_0)$ for a rectangular channel $d/a = 0,5$ with insulating walls, Barleon et al. (1995).

A good agreement between experiment and theory can be stated. The graph shows furthermore that the MHD-pressure losses increase with the channel aspect ratio d/a and approach the solution for Hartmann flow as $d/a \rightarrow \infty$. This indicates the dominating rolc of the Hartmann boundary layers as a source for electromagnetic pressure losses originating from Joule-dissipation.

An alternative to this representation is frequently used by introducing a dimensionless pressure gradient $\nabla p'$ of the form

$$\nabla p' = \frac{\Delta p/L}{\sigma v_0 B_0^2} \qquad (95)$$

where $\Delta p/L$ is the measured pressure gradient along the channel length L. In figure 17b this dimensionless pressure gradient is displayed as a function of the Hartmann number and the dimensionless so called interaction parameter N which is defined as

$$N = \frac{a\sigma B_0^2}{\rho u_0}. \qquad (96)$$

The graph contains measured data of Barleon et al. (1995) obtained in a rectangular channel with insulating walls and calculated values according to the outlined theory. There is a good agreement between the measured and the predicted pressure losses. Moreover, it is clearly seen that the measured pressure losses are independent of the interaction parameter. This means that the pressure losses are inversely proportional to the volumetric flux, giving the mean channel velocity as used to build the interaction parameter. The MHD-channel flow behaves thus in a wide range of Hartmann numbers considered like a laminar hydrodynamic flow with $\Delta p \sim 1/v_o$.

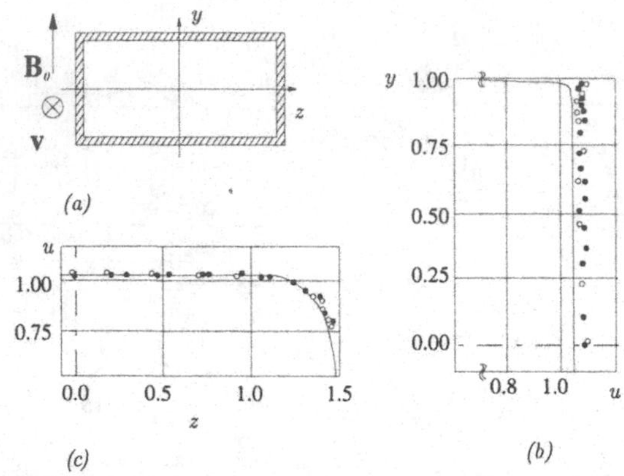

Figure 16. Sketch of the test channel of ? (?); the channel walls are insulating, aspect ratio $d/a = 1.5$, Hartmann number $Ha = 174$; $Re = 25300$ (·);$Re = 40200$ (∘), velocity profiles b) $u(y, z = 0,75)$, c) $u(y = 0.5, z)$.

? (?) have also performed local velocity measurements using Pitot tube probes. Their test channel had an aspect ratio $d/a = 1.5$.

Figure 16 shows normalized measured local velocity values and corresponding calculated values. While the measurements can not resolve the Hartmann boundary layers the side layers are clearly identified by the measurements. As expected the velocity profile is flat in the channel center even at these moderate Hartmann numbers.

Perfectly conducting walls For solving the equations (68a,b) for the case of perfectly conducting walls the boundary condition for the induced magnetic field change to a vanishing normal derivative at the wall $\frac{\partial b}{\partial n}(y, z = \pm d) = \frac{\partial b}{\partial n}(y = \pm 1, z) = 0$.

The procedure to solve the problem is completely analog to the one outlined in the previous chapter (see also Huges and Young (1966)). To demonstrate the differences of this case with $\sigma_w \to \infty$ compared to the case of insulating walls $\sigma_w = 0$ (see figure ??) the velocity and induced magnetic field isolines are shown in figure 3.

The velocity isolines demonstrate, that the area of the flat one velocity distribution is more extended for perfectly conducting walls. Areas of overspeed can be seen near the side walls. The current stream lines are nearly parallel to the z-axis except in the channel corner regions.

In figure 17 the calculated normalized pressure gradient is plotted versus the Hartmann number for the limiting cases insulating ($\sigma_w = 0$) and perfectly conducting ($\sigma_w \to \infty$) walls. It is clearly seen that perfectly conducting walls result in significantly higher MHD-pressure losses.

3 Approximate solutions for Ha≫1

In many technical applications liquid metal occurs as the working fluid in MHD-flow. This is usually so in applications in metallurgical processing or in nuclear power plants, where the

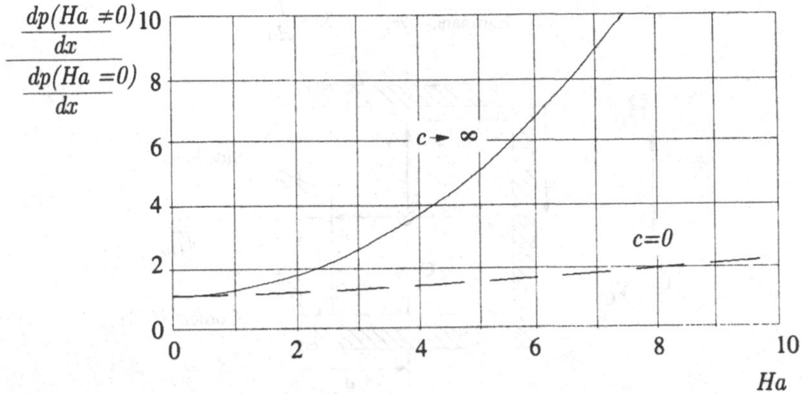

Figure 17. Calculated pressure gradient in MHD-flow in a rectangular channel normalised by the pressure gradient for laminar viscous channel flow; (-) perfectly conducting walls, (—) insulating walls.

iquid metal is used as a coolant. Liquid metals have typically an electrical conductivity of $\sigma \sim O(10^6)\,[A/(Vm)]$. Their kinematic viscosity is of the same order as that of water $\nu \sim O(10^{-6})\,[m^2/s]$. This gives rise to high Hartmann numbers of $Ha \sim O(10^3)$ even for relatively small magnetic field intensities of the Order $B \sim O(10^{-1})[A/m]$ and small geometrical dimensions $a \sim 10^{-1}m$. In this range of Hartmann numbers the analytical solutions are difficult to evaluate and numerical solutions require large CPU-times. Approximate solutions for $Ha \gg 1$, often called *core flow approximation*, are therefore extremely valuable.

3.1 The basic idea, y-symmetric cross sections

Similar to the formation of viscous boundary layers in high Reynolds number flow in high Hartmann number MHD-flow a distinct formation of magnetohydrodynamic boundary layers occur. This phenomenon has been identified e.g. in section 2.1 from special analytical solutions. For obtaining approximate solutions the flow domain is subdivided in a "core region" and "boundary layer regions" (and if necessary in "corner regions" if the wall surface is not smooth). For each region a simplified set of conservation equation holds, which may be solved analytically. The solutions are adjusted to the wall boundary conditions and are matched to each other at the domain boundaries. To be more specific we consider for example the flow in a rectangular channel as sketched in figure 18.

We distinguish the following subdomains: the core (subscript c); the Hartmann boundary layers (subscript H); the side wall boundary layer (subscript S) and the corner region. It is shown in a number of references that the corners do not contribute to the mass flux or pressure drop at leading order of approximation. Therefore these regions are not considered in more detail during this subsection.

We will construct an approximate solution of the boundary value problem for channel flow (see equations 11, 12) as a superposition of solutions for each subdomain. For instance the ve-

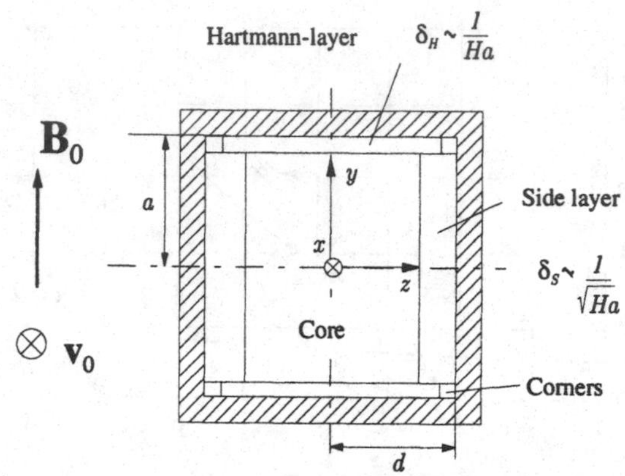

Figure 18. Channel geometry and the orientation of the magnetic field

locity distribution is represented by

$$u\,(y,z) = u_C + u_H + u_S. \tag{97}$$

In order to obtain simplified MHD-channel equations for the core region it is useful to rescale u and b in equations (11, 12) by $u \to Ha^{-1}\, u$ and $b \to Ha^{-1}b$. We obtain the following form of equations:

$$\frac{\partial b}{\partial y} + \frac{1}{Ha}\left(\frac{\partial^2 u}{\partial y^2} + \frac{\partial^2 u}{\partial z^2}\right) = -1, \tag{98}$$

$$\frac{\partial u}{\partial y} + \frac{1}{Ha}\left(\frac{\partial^2 b}{\partial y^2} + \frac{\partial^2 b}{\partial z^2}\right) = 0. \tag{99}$$

Next we have to provide solutions for the subdomains.

The core: For $Ha \to \infty$ the equations governing the flow in the core take the simplified form

$$\frac{\partial b_c}{\partial y} = -1, \quad \frac{\partial u_c}{\partial y} = 0. \tag{100}$$

The first equation just describes the asymptotic balance of pressure and Lorentz forces in the core of the channel. The second equation states that the velocity profile is "flat" with regards to a variation in direction of the **B**-field. (This is very general and holds for any y-symmetric cross section)

The core equations can be solved with the result that

$$b_c = -y, \quad u_c = u_c\,(z)\,, \tag{101}$$

where the constant of integration for b_c vanishes in the case of symmetry of the problem with respect to the z-axis.

The Hartmann layers: Next we discuss the behavior of the velocity and the induced magnetic field in the Hartmann boundary layer. By scaling the coordinate y with the normalized thickness of the Hartmann layer $\delta_H = Ha^{-1}$ and neglecting higher order terms we arrive at boundary layer equations in the form of ordinary differential equations for the variables u_H and b_H. As an example, consider the Hartmann layer near the upper wall of a channel at $y = Y$. For parallel Hartmann walls we have $Y = 1$.

With the Hartmann layer coordinate $\eta = Ha\,(y - Y)$ one arrives at a set of ordinary differential equations

$$\frac{\partial b_H}{\partial \eta} + \frac{\partial^2 u_H}{\partial \eta^2} = 0, \tag{102}$$

$$\frac{\partial u_H}{\partial \eta} + \frac{\partial^2 b_H}{\partial \eta^2} = 0. \tag{103}$$

The solutions that match smoothly the core solution $u_H, b_H \to 0$ at some distance from the wall as $\eta \to -\infty$ and which satisfy the boundary conditions at the wall are

$$u_H = -u_c \exp(\eta), \quad b_H = u_c \exp(\eta). \tag{104}$$

By applying the Shercliff thin wall condition at the Hartmann wall, at $y = Y, \eta = 0$,

$$n_y \frac{\partial}{\partial y}(b_H + b_c) + n_z \frac{\partial}{\partial z}(b_H + b_c) + \frac{1}{c}(b_H + b_c) = 0 \tag{105}$$

and neglecting $n_z \frac{\partial}{\partial z}(b_H + b_c)$ one arrives with

$$n_y (Ha\, u_c - 1) + \frac{1}{c}(u_c - Y) = 0, \tag{106}$$

at

$$u_c(z) = \frac{n_y c + Y}{n_y c Ha + 1}, \tag{107}$$

a relation that is valid for any y-symmetric duct cross section. The unit normal to the wall is $\hat{\mathbf{n}} = (0, n_y, n_z)$. For the case of parallel Hartmann walls with $n_y = 1, Y = 1$, one recovers the asymptotic representation of Hartmann flow obtained by Chang and Lundgren (1961). Note, the matching of the Hartmann layer solution with the core solution provides now the value of core velocity that was undetermined up to now. The assumption that the second term in equation (105) is negligible holds for almost the whole duct cross section. It becomes, however, invalid in regions where n_y is small. To be more precise, it fails, when $n_y Ha\ u_c \gg n_z \frac{\partial}{\partial z} u_c$ is not guarantied. The relationship for u_c fails if the channel walls are either parallel or nearly parallel to the external magnetic field. In that case the derivatives $Y' = \partial Y/\partial z$ in $n_y = \left(1 + Y'^2\right)^{-1/2}$ give rise to a singular behavior and detailed analysis is required. This is the case in regions where the wall becomes tangential to the magnetic field. In circular duct cross sections such regions do not contribute to the pressure drop and flow rate at leading order of approximation. The result for the core velocity shows the functional dependency of this quantity on the Hartmann number, the wall conductance parameter the local channel height and the local wall inclination.

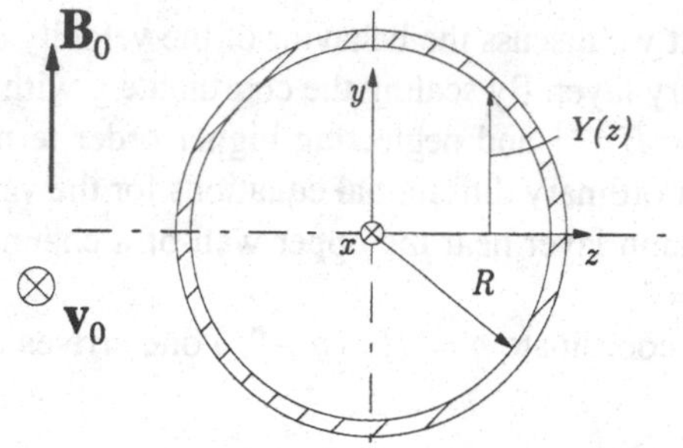

Figure 19. Geometry and coordinates in a circular pipe

For the case of $c = 0$ the restriction on $\hat{\mathbf{n}}$ are much weaker and one finds $u_c = Y$. The local velocity is proportional to the channel local height. As an example see the flow in a circular pipe in figure 20.

For perfectly conducting walls $c \to \infty$ the core velocity becomes independent of the walls contour, $u_c = Ha^{-1}$.

Combining the solutions for the core region and the Hartmann boundary layer we get the following representation of the variables:

$$u = u_c + u_H = u_c(z)\{1 - \exp[Ha(|y| - 1)]\},$$
$$b = b_c + b_H = -y \pm u_c(z)\exp[Ha(|y| - 1)] \quad \text{for } y \gtrless 0. \tag{108}$$

It can be demonstrated that the assumption of the symmetry of the channel contour is not necessary for the case of insulating channel walls. In this case the core velocity is given by the equation

$$u_c = \frac{1}{2}(Y_{top} - Y_{bot}). \tag{109}$$

where $Y_{top} - Y_{bot}$ is the distance between the upper and lower channel contour.

3.2 Circular ducts

The outlined general calculation method for channel flow at high Hartmann numbers is now applied to a circular duct geometry. The situation is depicted in figure 19. We have for the duct wall $Y = (1 + z^2)^{1/2}$ and here $n_y = Y$. The duct radius $r = a$ has been taken as the geometrical scale.

A straight forward evaluation gives for the core velocity distribution

$$u_c(z) = \frac{c + 1}{cHa + \frac{1}{Y}} \tag{110}$$

This solution differs from the one obtained by Shercliff (1954) by the factor cHa as he considers the case $c << 1$. By integrating $u_c Y$ in the interval$-1 \le z \le 1$ we obtain the volumetric flow rate. It is convenient to use the cylindrical coordinates for the integration

$$Q = 4\frac{c+1}{Ha}\int_0^{\pi/2} \frac{\cos^3(\tau)}{1 + cHa\cos(\tau)}d\tau. \tag{111}$$

The result finally reads

$$Q = 4\frac{c+1}{Ha}\left[\frac{\pi}{4cHa} - \frac{1}{(cHa)^2} + \frac{\pi}{2(cHa)^3} - \frac{2}{(cHa)^3}\frac{\operatorname{arctanh}\left(\sqrt{\frac{cHa-1}{cHa+1}}\right)}{\sqrt{(cHa)^2-1}}\right]. \tag{112}$$

Chang and Lundgren (1961) show that the expression for Q does not have a singularity as $cHa \to 0$ since for that case the flow rate asymptotes to

$$Q = 4\frac{c+1}{Ha}\left[\frac{2}{3} - \frac{3\pi}{16}cHa + ...\right] \quad \text{for } cHa \ll 1. \tag{113}$$

In figure 20 the core velocity distribution along the z-axis is presented for different parameter combinations cHa. It is seen that the velocity profile flattens with increasing values of the product cHa. An exact solution for a circular duct flow with insulating duct walls has been derived by Gold (1962). An improved approximation of the presented core flow solution has been elaborated by Shercliff (1962).

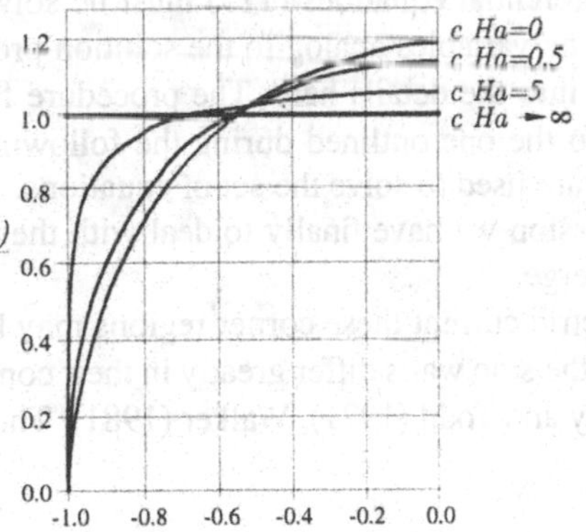

Figure 20. Core velocity in a circular pipe flow for different wall conductivity. The velocity is normalized with the average velocity v_0.

The solution shown so far is valid except near the sides, where the magnetic field becomes parallel with the wall. These areas have been examined by Roberts (1967) and are known as "Roberts layers". As far as the overall volumetric flow rate in the channel is concerned these regions do not significantly contribute to it as the area scales as $Ha^{-2/3} * Ha^{-1/3}$ measured normal and tangential to the wall. Moreover, the velocity profile drops off to a zero value faster near the side than in any other region. A correction of the flow rate due to the Roberts layers would be on the order of $Ha^{-7/3}$.

3.3 Rectangular ducts

The solution for core velocity according to equation (108) holds for any duct shape, especially for the core of rectangular ducts where $n_y = Y = 1$. The solution breaks down, however, near the side walls, where n_y vanishes or to be more precise, in a distance $O\left(Ha^{-1/2}\right)$ from the side walls. In the region of the side wall boundary layers we rescale the coordinates by stretching z with the boundary layer thickness

$$\delta_s \sim Ha^{-1/2}. \tag{114}$$

With the side layer coordinate $\zeta = Ha^{1/2}\,(z-d)$ one arrives at

$$\begin{aligned} \frac{\partial b_s}{\partial y} + \frac{\partial^2 u_s}{\partial \zeta^2} &= 0, \\ \frac{\partial u_s}{\partial y} + \frac{\partial^2 b_s}{\partial \zeta^2} &= 0. \end{aligned} \tag{115}$$

It is to be noticed here that this thickness is certainly different in its dependence on Ha and much larger than that of the Hartmann layer. The balancing of the viscous and the electromagnetic terms confirms that the scale chosen for a side layer approximation is correct.

If more details about the distribution of u_s and b_s are wanted in the boundary layer range the set of parabolic partial differential equations (115) must be solved. This can be done e.g. by a Fourier decomposition of the variables analog to the solution procedure for the heat conduction equation. We shall not go into the details here. The procedure for the case of insulating walls, however, may be analog to the one outlined during the following subsection, where similarity transformation techniques are used to solve the set of equations.

To complete our discussion we have finally to deal with the corner regions where the Hartmann and the side layer merge.

With regard to the electric current these corner regions may become of significant influence if the Hartmann walls and the side walls differ greatly in their conductivity. Detailed descriptions are given by e.g. Temperley and Todd (1971), Walker (1981). The role of the corner is addressed by Tabeling (1982).

4 Free shear layers parallel to the B-field in two-dimensional channel flow

Free shear layers can be generated in MHD-channel flow by strong local inhomogeneities of wall boundary conditions. The origin of these inhomogeneities, frequently called singularities, may be of electrical, geometrical or material nature such as local sources of electrical current, corners, or discontinuities of the electrical wall conductivity. In the following some cases are considered.

4.1 Electrodes

For our discussions we consider a plate channel with insulating walls. In one of the walls the current **I** is injected from a line electrode. The same amount of current is extracted at the other wall, again by a line electrode. The most simple case is when both electrodes lie on the same magnetic field line. The situation is sketched in figure 21. More complicated situations arise when the electrodes are shifted by some distance as shown in figure 23. Both cases have been considered by Hunt and Williams (1968). Here the main ideas are outlined that lead to their solution.

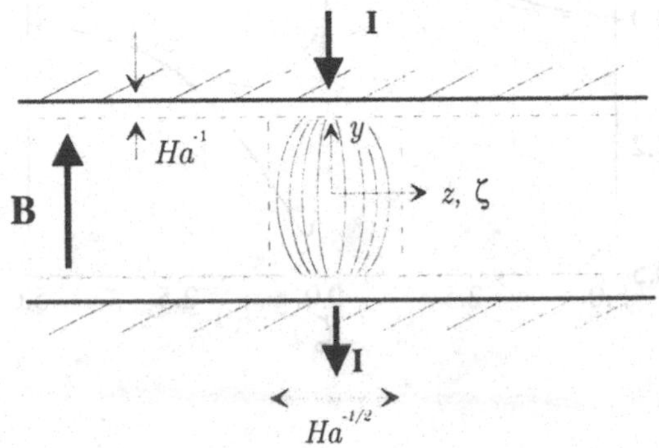

Figure 21. Injection of current by line electrodes. The current enters the fluid at $y = 1$ and leaves the fluid at $y = -1$. The fluid region is quiescent everywhere, except within a thin region near $z = 0$, where shear layers of width $Ha^{-1/2}$ spread along magnetic field lines.

The governing equations are essentially the same as those presented earlier (equations 11, 12). If there is no driving pressure gradient the right-hand side of the momentum equation vanishes. With the so-called Elsasser variables

$$X = u + b, \quad Y = u - b \tag{116}$$

Hunt and Williams (1968) use the governing equations in the form

$$\frac{\partial^2 X}{\partial y^2} + \frac{\partial^2 X}{\partial z^2} + Ha\frac{\partial X}{\partial y} = 0, \tag{117}$$

$$\frac{\partial^2 Y}{\partial y^2} + \frac{\partial^2 Y}{\partial z^2} - Ha\frac{\partial Y}{\partial y} = 0. \tag{118}$$

Now one sees the big advantage of the new variables. The equations decouple in the unknowns, a fact that simplifies considerably the analysis.

At both walls at $y = \pm 1$ there is no slip, and the injected current leads to a jump of the magnetic field b near the electrodes

$$u = 0, \quad b = \pm 1 \quad \text{for } z \lessgtr 0. \tag{119}$$

This requires at the walls e.g. for X

$$X = \pm 1 \quad \text{for } z \lessgtr 0. \tag{120}$$

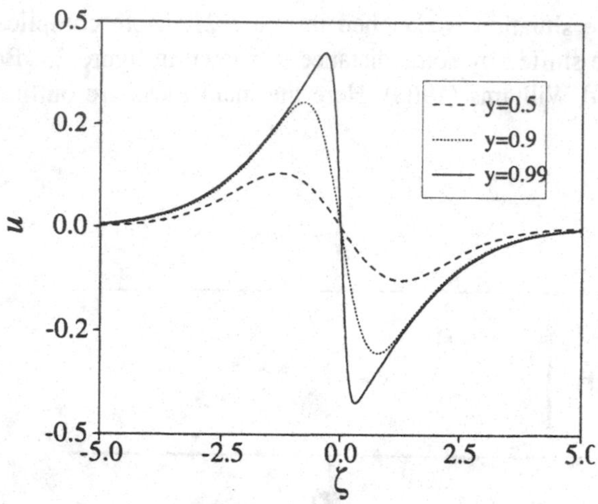

Figure 22. Velocity profile for aligned line electrodes. The velocity vanishes at the symmetry line $y = 0$ and shows highest values when approaching the Hartmann layers.

By introducing the relevant scales in form of the stretched coordinates $\zeta = \sqrt{Ha}z$ one finds at leading order

$$\frac{\partial^2 X}{\partial \zeta^2} + \frac{\partial X}{\partial y} = 0, \tag{121}$$

an equation that permits self-similar solutions of the form

$$X = -\operatorname{erf}\left(\frac{\zeta}{2\sqrt{1-y}}\right). \tag{122}$$

After a corresponding solution for Y has been obtained one can evaluate the velocity as $u = \frac{1}{2}(X + Y)$ and find

$$u = \frac{1}{2}\left\{\operatorname{erf}\left(\frac{\zeta}{2\sqrt{1+y}}\right) - \operatorname{erf}\left(\frac{\zeta}{2\sqrt{1-y}}\right)\right\}. \tag{123}$$

This result holds, except within the Hartmann layers where additional viscous corrections are required to satisfy no-slip. The details are given by Hunt and Williams (1968) and are not repeated here. Instead the physical aspects are outlined that lead to the solution as shown above. Results valid outside the Hartmann layers are displayed in figure 22 as a function of the stretched scale $\zeta = \sqrt{Ha}z$. The velocity vanishes at the symmetry line $y = 0$ and shows highest values when approaching the Hartmann layers. For negative values of y the sign in velocity is reversed.

The solution is created by two perturbations that travel along field lines and diffuse in transverse direction. One of these is X, traveling downward, the other is Y traveling upward. The perturbations spread in z-direction and exhibit the self similar profiles as shown in equation (122). The variable X has its physical source at $y = 1$. Consequently X has no Hartmann layer

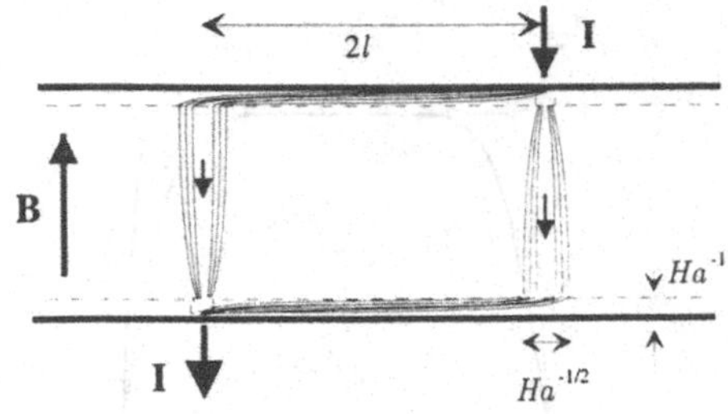

Figure 23. Injection of current by line electrodes. The current enters the fluid at $y = 1$ and leaves the fluid at $y = -1$. Near $z = l$ a thin layer spreads from the top towards the bottom, while near $z = -l$ the layer spreads from the bottom to the top. The parallel layers have typical thickness as $Ha^{-1/2}$.

at $y = 1$, but one near the other wall at $y = -1$. The same arguments hold for the upward travelling quantity Y, with source at $y = -1$ and Hartmann layer at $y = 1$.

An interesting situation arrises when the two electrodes are shifted in the horizontal direction. Suppose the upper electrode is now at $z = l$, while the lower one is at $z = -l$ as shown in figure 23. The current that is injected at the top splits into two parts. One part in conducted by the top Hartmann layer, while the other one finds its path via the internal layer to the bottom Hartmann wall. Near the bottom electrode the currents are again collected. This gives rise to a second parallel layer at the position of the electrode. The result is that everywhere in the fluid the current density vanishes, except in the parallel layers and in the intermediate part of the Hartmann layers.

The current flux in the Hartmann layers creates there a uniform potential gradient or equivalently an electric field that is imposed to the fluid in the central core. According to Ohm's law the flow in the interior core is driven as

$$u = -E_z = 1. \tag{124}$$

The profile of the core velocity may be seen from figure 24. The lateral gradient is more expressed near the position of the electrodes, which has its reason in the fact that the disturbances spread from these points and diffuse at larger distances.

4.2 Discontinuous wall conductivity

An analogous effect can also be observed if instead of a forced current a discontinuity of the wall conductivity exists. This gives rise to non uniform potential gradient (electric field) in the outer region of the Hartmann boundary layer which in turn result in a local velocity wake flow. Disregarding the details inside the spreading parallel layers one can deduce the velocity field from the solution given by Chang and Lundgren (1961) (see also equation (55))

$$u = \frac{c+1}{c\,Ha+1}, \tag{125}$$

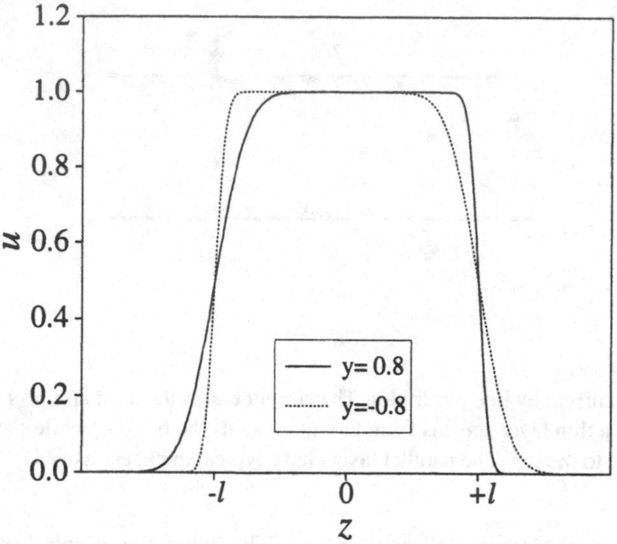

Figure 24. Velocity in the core between displaced electrodes

which leads to velocities of order unity in region where $c \ll Ha^{-1}$, while the velocity becomes negligibly small, $O\left(Ha^{-1}\right)$, in regions where the wall conductivity is high

$$u = \begin{cases} 1 & \text{for} \quad c = 0 \\ Ha^{-1} & \qquad c = \infty \end{cases} . \tag{126}$$

An experimental demonstration of a free shear layer in a MHD-channel flow performed by Rosant (1976) can be seen in figure 25. In a rectangular straight channel with partly insulating and partly highly conducting walls the velocity distribution along the axis of symmetry parallel to the Hartmann walls shows a sharp decline near the channel center line which connects the locations of the wall conductance discontinuity. The graph show also the increase of the velocity near the side wall because of the formation of side wall boundary layers with their typical "jet" characteristic.

4.3 Duct with corners

The general relationship for the core-velocity distribution equations (107) shows that corner i.e. discontinuities in derivative of the wall contour function may lead to a discontinuity in the core velocity. This indicates the existence of a shear region spreading along magnetic field lines from the discontinuity (here the corner) into the fluid. Indeed the locally non homogeneous current distribution at wall corners leads at very high Hartmann numbers to a strong variation of the velocity across a magnetic field line running through the particular corner. The situation is depicted in Figure 26.

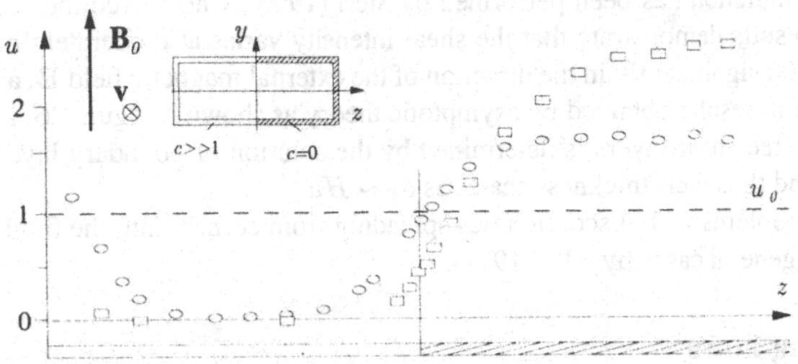

Figure 25. Experimental results for core velocity in a duct with partly insulating and partly conducting walls performed by Rosant (1976). (o) $Ha = 90$, (□) $Ha = 270$.

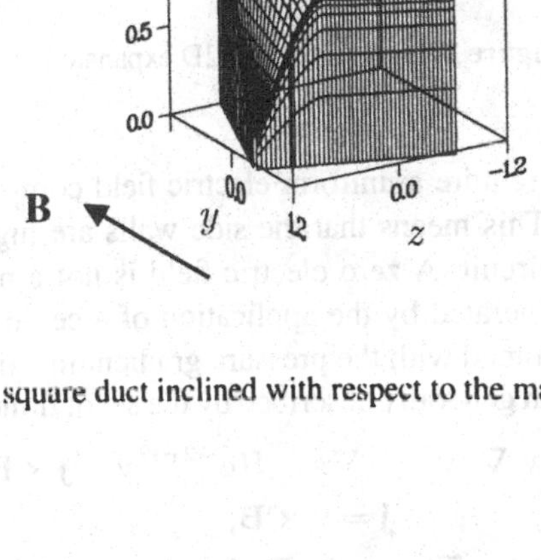

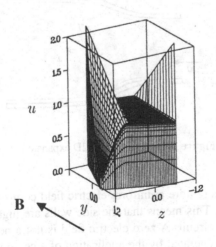

Figure 26. Velocity profile in a square duct inclined with respect to the magnetic field by 20°. $Ha = 10^4$, $c = 0.1$ (Bühler (1994)).

If in addition to the geometrical singularity by a corner a material singularity by adjacent different wall conductivities may occur at the same corner point and enhance the effects. A full numerical simulation has been performed by Sterl (1990), who solved the full set of MHD equations. His results demonstrate that the shear intensity varies at moderately high Hartmann number $Ha \simeq 200$ significantly in the direction of the external magnetic field $\mathbf{B}$, a point that can be observed also in results obtained by asymptotic theory as shown in figure 26. In any case the thickness of the free shear layers is determined by the solution of boundary layer equations of parabolic type and thus their thickness scales as $\delta_s \sim Ha^{-1/2}$.

MHD flow problems with discontinuities spreading from corners into the fluid has been discussed for more general cases by Alty (1971).

5 Developing flows

5.1 Flow in duct expansions for $Ha \gg 1$

Next we shall analyze formally the inertia effects on MHD-channel flow. As a typical hydraulic component, where inertia forces may dominate the flow, we choose a two-dimensional duct expansion as sketched in figure 27. The location of the two-dimensional expansion walls in an (x, y)-coordinate system may be described by the contour functions $F_t(x)$ and $F_b(x)$. The external magnetic field is aligned to the y-axis. There are Hartmann boundary layers along the channel walls. The depth of the expansion in z-direction is assumed to be large enough to assure a two-dimensional flow characteristic.

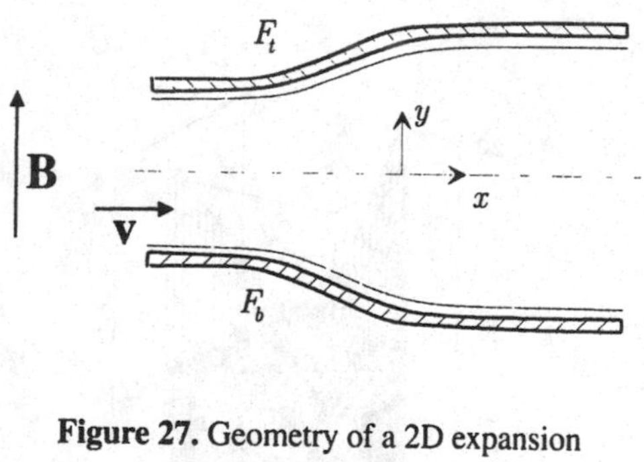

Figure 27. Geometry of a 2D expansion

Two dimensional solutions require a uniform electric field component in z-direction which is taken here to zero, $\mathbf{E} = 0$. This means that the side walls are highly conducting electrodes that are connected by a short circuit. A zero electric field is not a necessary assumption. Any constant electric field that is generated by the application of a certain voltage to the side walls can be admitted and easily combined with the pressure gradient in axial direction.

The steady MHD-flow is than generally described by the set of dimensionless MHD-equations

$$N^{-1}(\mathbf{v}\cdot\nabla)\mathbf{v} = -\nabla p + Ha^{-2}\nabla^2\mathbf{v} + \mathbf{j}\times\mathbf{B} \tag{127}$$

$$\mathbf{j} = \mathbf{v}\times\mathbf{B}, \tag{128}$$

$$\nabla\cdot\mathbf{v} = 0,\ \nabla\cdot\mathbf{j} = 0. \tag{129}$$

Note, the scale of variables is different here than in that previously introduced. The motivation behind this is that the flow rate is considered now as a main parameter and the average velocity in some cross section is used as velocity scale v_0 while the magnetic field is scaled by B_0.

For two-dimensional flow we have the following set of variables and dependencies:

$$\mathbf{v}(x,y) = (u, v, 0), \quad \mathbf{j}(x,y) = (j_x, j_y, j_z), \quad \mathbf{B} = (0,1,0), \quad p(x,y) \;. \tag{130}$$

With Ohm's law

$$j_z = u, \tag{131}$$

we find the two-dimensional equations for momentum

$$\frac{1}{N}\left(u\frac{\partial}{\partial x} + v\frac{\partial}{\partial y}\right)u = -\frac{\partial p}{\partial x} + \frac{1}{Ha^2}\left(\frac{\partial^2}{\partial x^2} + \frac{\partial^2}{\partial y^2}\right)u - u, \tag{132}$$

$$\frac{1}{N}\left(u\frac{\partial}{\partial x} + v\frac{\partial}{\partial y}\right)v = -\frac{\partial p}{\partial y} + \frac{1}{Ha^2}\left(\frac{\partial^2}{\partial x^2} + \frac{\partial^2}{\partial y^2}\right)v. \tag{133}$$

Generally, these equations cannot be solved analytically in closed form. Either a numerical approach has to be taken or approximate solutions in asymptotic form must be derived. For many technical applications asymptotic solutions for high Hartmann-numbers $Ha \gg 1$ and interaction parameters $N \gg 1$ can indeed be obtained. Such an approach will be discussed next.

5.2 Expansion flow with $Ha \gg 1$ and $N \gg 1$

The core flow approximation provides a composite solution consisting of an analytic expression for the flow in the core region of the domain to which a boundary layer solution, i.e. in our case an asymptotic form of a Hartmann-solution is matched. For the core flow a solution can be constructed in a similar way like the one for duct flow with general duct cross-section. The simplified form of the momentum equations (132, 133) read

$$\frac{\partial p_c}{\partial x} = -u_c, \quad \frac{\partial p_c}{\partial y} = 0. \tag{134}$$

Here the subscript c denotes the core region.

From these equations we conclude

$$\frac{\partial p_c}{\partial x} = -u_c(x) \,. \tag{135}$$

Conservation of mass requires that $u\,(F_t - F_b) = const \doteqdot 1$ and the core velocity becomes

$$u_c = (F_t - F_b)^{-1} \,. \tag{136}$$

The MHD inertialess pressure loss can be calculated by integrating (135) with (136).

$$p_c(x_2) - p_c(x_1) = -\int_{x_1}^{x_2} (F_t - F_b)^{-1}\, dx. \tag{137}$$

The derived core flow approximation can only be valid as long as the core velocities and their derivatives are much smaller than $O(N)$; otherwise the convective transport terms have to be taken into account. This may typically occur in duct expansions with sharp bends. Hunt and Leibovich (1967) have shown that for high Hartmann numbers generally free shear layers originate from sharp bends or edges in a duct expansion. The shear layer is in the direction of the external magnetic field and may give rise to a significant flow redistribution and contribute to the pressure drop.

5.3 The formation of inertia controlled free shear layers

We shall next discuss some characteristic features of free shear layers originating e.g. from sharp bends in ducts. These shear layers are called Ludford layers as they were first identified by Ludford (1960) who investigated the flow around obstacles in channels. The shear layers may be governed by the balance of inertia, Lorentz and viscous forces and accordingly their thickness may be controlled by the Hartmann number and the interaction parameter. In duct flows similar types of inertial layers are possible that originate from sharp corners in expansions. A schematic view of a duct expansion with a local shear layer spreading from the sharp wall bend to the opposite wall is seen in figure 28. In the previous section an equation (136) for the core velocity has been derived that leads to continuous axial velocities even across internal layers. Nevertheless, the y-component of velocity is discontinuous near the edge since the wall normal direction used in the boundary condition $\mathbf{n} \cdot \mathbf{v} = 0$ changes abruptly.

Conservation of mass requires

$$\frac{\partial v_c}{\partial y} = -\frac{\partial u}{\partial x} = -\frac{\partial}{\partial x}\left(F_t - F_b\right)^{-1}, \tag{138}$$

and, for our specific case with F_t =*const*, $v\left(y = F_t\right) = 0$, it follows

$$v_c = \left(F_t - y\right)\frac{\partial}{\partial x}\left(F_t - F_b\right)^{-1}. \tag{139}$$

The sudden occurrence of a vertical velocity component in the expanding section gives rise to a local shear flow. Within the vicinity of the sharp corner the core flow approximation fails and the complete form of the momentum transport equations (132, 133) must be considered in a domain δ, which is the shear layer thickness. Within this shear layer the discontinuity in the y-component of the core velocity must be bridged.

The dependency of δ on the interaction parameter N and/or the Hartmann can be derived from the basic equations by introducing the stretched coordinate

$$\xi = x/\delta. \tag{140}$$

Denoting the channel height at the edge position by $F_t(0) - F_b(0) = 1$, it is assumed that $\delta \ll 1$. For the horizontal velocity component in the shear layer an ansatz in the form

$$u = 1 + \delta U(\xi, y) \tag{141}$$

is made which accounts for the core flow solution and a correction in the shear layer. Introducing the stretched coordinate ξ into the continuity and momentum equations and neglecting inertia

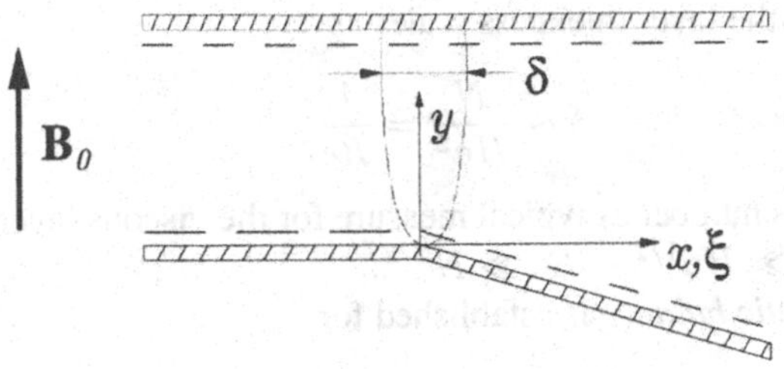

Figure 28. Geometry and coordinates at a sharp corner as considered by Hunt and Leibovich (1967)

and viscous terms of the order $O(\delta)$ compared to terms of $O(1)$, we obtain the following set of simplified equations which govern the flow in the shear layer region:

$$\frac{\partial U}{\partial \xi}+\frac{\partial v}{\partial y}=0, \tag{142}$$

$$\frac{1}{N}\frac{\partial U}{\partial \xi}=-\frac{1}{\delta}\frac{\partial p}{\partial \xi}-1-\delta U+\frac{1}{Ha^2}\frac{1}{\delta^2}\frac{\partial^2 U}{\partial \xi^2}, \tag{143}$$

$$\frac{1}{N}\frac{1}{\delta}\frac{\partial v}{\partial \xi}=-\frac{\partial p}{\partial y}+\frac{1}{Ha^2}\frac{1}{\delta^2}\frac{\partial^2 v}{\partial \xi^2}. \tag{144}$$

Assuming that $\delta \geq O\left(Ha^{-1}\right)$ and $N >> 1$, $Ha >> 1$ we can simplify the equations further to yield

$$\frac{1}{\delta}\frac{\partial p}{\partial \xi}=-1-\delta U\left(\xi,y\right). \tag{145}$$

This relation indicates that the pressure loss across the shear layer is essentially of the order $O(\delta)$. Using equation (145) and the continuity equation (142) the pressure term in the second momentum equation (144) can be eliminated together with the ξ-component U of the velocity. We obtain a fourth order partial differential equation for the longitudinal velocity $v(\xi,y)$ in the shear layer in the form

$$\underbrace{\frac{1}{N}\frac{1}{\delta}\frac{\partial^3 v}{\partial \xi^3}}_{\text{inertia}}=\underbrace{\frac{1}{Ha^2}\frac{1}{\delta^2}\frac{\partial^4 v}{\partial \xi^4}}_{\text{viscous}}\;\underbrace{-\delta^2\frac{\partial^2 v}{\partial y^2}}_{\text{electromagnetic}} \tag{146}$$

This equation represents a force balance in the shear layer between inertia, viscous, and Lorentz forces.

Consider the case when, depending on the ratio of the prefactors of the different terms in equation (146) two of the three acting forces balance in the first place. Accordingly the shear layer thickness and the velocity may scale in different powers of the dimensionless groups Ha and N.

A *viscous-inertia balance* requires

$$\frac{1}{N}\frac{1}{\delta}\sim\frac{1}{Ha^2}\frac{1}{\delta^2}\gg\delta^2. \tag{147}$$

This leads to the result that the layer thickness scales as

$$\delta \sim \frac{N}{Ha^2} = \frac{1}{Re} \tag{148}$$

with the hydraulic Reynolds number as typical measure for the viscous layer. This result holds if Re is high enough, for $Re \gg Ha^{1/2}$, $Re \gg N^{1/3}$.

A *viscous-electromagnetic balance* is established for

$$\frac{1}{Ha^2}\frac{1}{\delta^2} \sim \delta^2 \gg \frac{1}{N}\frac{1}{\delta}, \tag{149}$$

with thickness of the layer of

$$\delta \sim Ha^{-1/2}, \tag{150}$$

if $N \gg Ha^{3/2}$.

Finally one can deduce that for a *inertia-electromagnetic balance* the relation

$$\frac{1}{N}\frac{1}{\delta} \sim \delta^2 \gg \frac{1}{Ha^2}\frac{1}{\delta^2} \tag{151}$$

must hold that determines the layer thickness as

$$\delta \sim N^{-1/3}, \tag{152}$$

for $N \ll Ha^{3/2}$. It has been shown that, depending on the combination of the parameters Ha and N, one can have layers of different typical thickness. The pressure loss across the shear layer then depends essentially on this parameter and using equation (145) one finds

$$\Delta p \sim Ha^{-1/2} \quad \text{or} \quad \Delta p \sim N^{-1/3} \tag{153}$$

for $N \gg Ha^{3/2}$ or $N \ll Ha^{3/2}$, respectively.

5.4 Solution for the inertia controlled shear layer

For the specific single sided expansion sketched in figure 28 a solution for v is discussed along the lines given by Hunt and Leibovich (1967). Assuming the "Ludford layer" is inertia controlled as the control parameter Ha is in the range $\sqrt{Re} << Ha << Re^2$ than within the layer the differential equation

$$\frac{\partial^3 v}{\partial \xi^3} + \delta^3 N \frac{\partial^2 v}{\partial y^2} = 0 \tag{154}$$

holds. The error made for neglecting the viscous effects is of the order $O\left(\left[Ha/Re^2\right]^{2/3}\right)$. With the proper scale for the layer thickness $\delta = N^{-1/3}$ one obtains the following form of the shear layer equation

$$\frac{\partial^3 v}{\partial \xi^3} + \frac{\partial^2 v}{\partial \eta^2} = 0. \tag{155}$$

The new coordinate along the magnetic field is obtained by

$$F_t - y = (F_t - F_b)(1 - \eta) \tag{156}$$

which gives near $x = 0$ with $(F_t(0) - F_b(0)) = 1$

$$F_t - y = 1 - \eta \tag{157}$$

A solution of this equation must be matched for $\xi \to \pm\infty$ to the core flow values. For our chosen geometry we have

$$v_c = 0 \quad \text{as} \quad \xi \to -\infty, \tag{158}$$

$$v_c = F_b'(F_t - y) \quad \text{as} \quad \xi \to +\infty. \tag{159}$$

Here, $F_b' = \partial F_b / \partial \xi$. The boundary conditions for the shear velocity v at the channel wall are

$$v(\xi, \eta = 1) = 0; \tag{160}$$

$$v(\xi, \eta = 0) = 0 \quad \text{for } \xi < 0 \tag{161}$$

$$v(\xi, \eta = 0) = F_b' \quad \text{for } \xi > 0. \tag{162}$$

An explicit solution has been given by Hunt and Leibovich (1967) using Fourier transform techniques

$$v = F_b' \frac{1}{2\pi i} \int_{-\infty}^{\infty} \frac{\sin\left[(i\omega)^{3/2}(1-\eta)\right]}{\sin(i\omega)^{3/2}} \exp(i\omega\xi) \frac{d\omega}{\omega}. \tag{163}$$

This integral has to be evaluated taking into account the matching conditions. This gives according to Hunt and Leibovich (1967)

$$\frac{v(\xi,\eta)}{F_b'} =$$

$$\begin{cases} (1-\eta) - \frac{4}{3}\sum_{n=1}^{\infty} \frac{(-1)^{n+1}}{n\pi} \exp\left(-\frac{(n\pi)^{2/3}}{2}\xi\right) \cos\left[\frac{\sqrt{3}}{2}(n\pi)^{2/3}\xi\right] \sin\left[n\pi(1-\eta)\right] & \text{for } \xi > 0; \\ \frac{2}{3}\sum_{n=1}^{\infty} \frac{(-1)^{n+1}}{n\pi} \exp\left[(n\pi)^{2/3}\xi\right] \sin\left[n\pi(1-\eta)\right] & \text{for } \xi < 0; \end{cases} \tag{164}$$

Note, both solutions match smoothly at $\xi = 0$ with a value of $v/F_b' = \frac{1}{3}(1-\eta)$. The solution is displayed in figure 29. The character of a spatially concentrated shear layer with steep velocity gradients in the vicinity of the edge location ($\xi = 0, \eta$) is obvious. There is even an excess velocity formed within the Ludford layer.

5.5 Bend flows

Ludford layers as described above occur similarly in bend geometries. One such case is the bend, which turns the flow from a direction initially perpendicular to the field into a direction perfectly

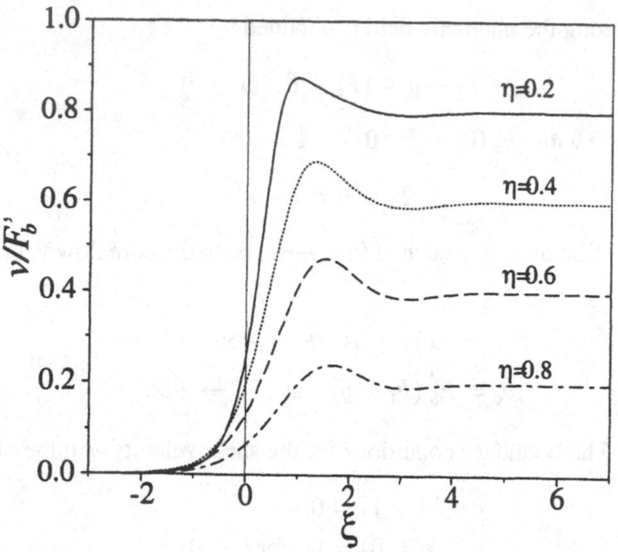

Figure 29. Calculated vertical velocity distribution in a Ludford shear layer generated by an abrupt change of a duct wall contour (see Hunt and Leibovich (1967)).

aligned with the field. Such a situation is sketched in figure 30. The figure shows the geometry of the bend. Because of symmetry with respect to the z-coordinate it is sufficient to show only one half of the geometry. Without going into the details of the analysis the main features of the flow are briefly described.

The flow in the part that is perpendicular too the field exhibits the boundary layers as they are known from duct flows. There are the Hartmann layers at the wall to which the magnetic field is normal. These layers are not explicitly shown in the sketch, since they do not contribute to the mass flux. Near the other wall to which the magnetic field is tangential one finds the side layers with $\delta_s \sim Ha^{-1/2}$, just as for straight duct flows.

When approaching the corner, say at $x = 0$, the fluid coming from the core I meets the internal Ludford layer. This layer distributes the flow among the core II and the parallel layers with flow rates Q_2, Q_3, in part II of the bend. The flow rates in the layers are plotted as a function of the vertical coordinate in figure 30. The most interesting thing is that for large values of y the flow rates in the layers Q_1 and Q_3 are larger than the total flow rate. This requires negative values of the flow rate Q_2. The core II does not contribute to the flow rate in that part of the bend, ($v = 0$) but redistributes the flow in planes $y = const$ among the different layers.

The pressure drop along the bend can be determined theoretically by an asymptotic theory valid for large Ha and N as outlined by Molokov and Bühler (1994). The theoretical predictions can be confirmed by experiments performed by Stieglitz et al. (1996) who find for large N, ($N^{-1/3} \to 0$) almost the same results. Moreover, the experiments give results over a wide range of the interaction parameter N ranging from about $10^2 < N < 10^5$ for which no theoretical results are available until present day. The results are shown in figure 31 as a function of $N^{-1/3}$

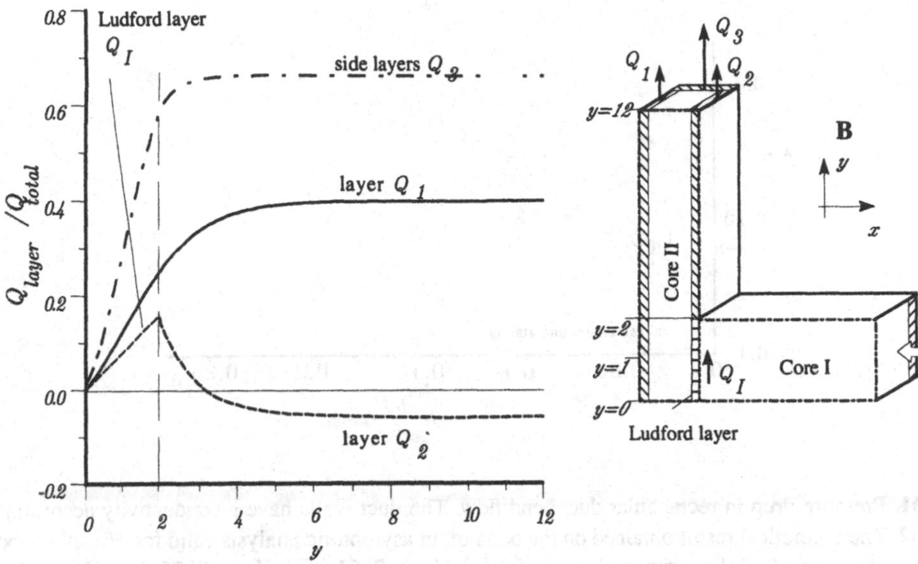

Figure 30. Bend flow: Flow rates carried by the different boundary layers and by the internal Ludford layer as given by Stieglitz et al. (1996).

to demonstrate the linear behavior of the pressure drop on this quantity. Such a result suggests now that the inertia part of pressure drop is essentially due to the interaction of the flow with the Ludford layer of thickness $\delta \sim N^{-1/3}$ that may result in pressure drop on this order of magnitude.

Stieglitz et al. (1996) give an engineering formula for the pressure drop caused by a rectangular duct bend flow that reads

$$\Delta p = \Delta p_{Ha,N\to\infty} + 0.406N^{-0.337} + 0.0934Ha^{-0.565}, \tag{165}$$

where $\Delta p_{Ha,N\to\infty}$ is the pressure drop in the asymptotic limit, when $Ha, N \to \infty$. The exponents and prefactors have been obtained by best fitting the experimental data and confirm thereby the behavior as expected for inertial-electromagnetic and viscous-electromagnetic interaction.

The role of the Ludford layers becomes most pronounced when the bend is turned by an angle to form a backward elbow as considered theoretically by Moon et al. (1991) or by Bühler (1994) (see figure 32). The experimental results that have been presented by Stieglitz et al. (1996) also for a backward elbow. Stieglitz (1999, unpublished; data extracted from the experiments described by Stieglitz et al. (1996)) tries to asses the thickness of the Ludford layer by measuring electric potential gradients along a traverse indicated in the sketch in figure 32. Near the internal layer the potential has two inflection points and it was the idea to use their distance as a qualitative measure for the layer thickness. The results indicate again that the layer thickness depends on inertia as $\delta \sim N^{-1/3}$.

In heat transfer applications for fusion reactor blankets it may occur that a number of bends like the one shown in figure 30, are sandwiched at common conducting side walls. For such situ-

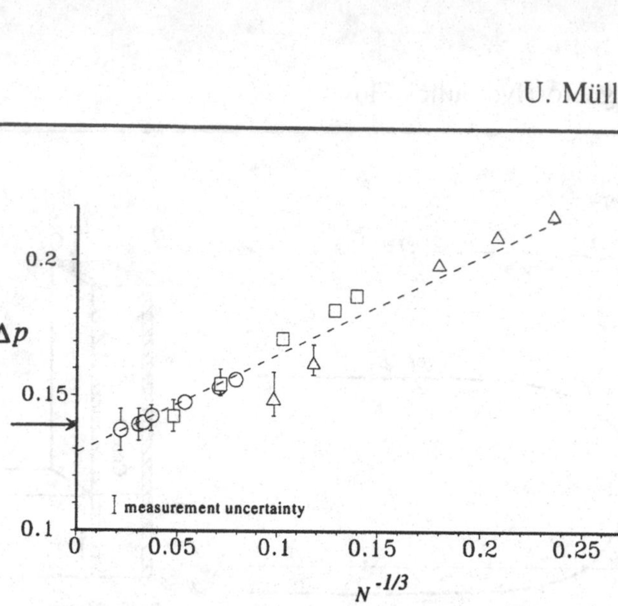

Figure 31. Pressure drop in rectangular duct bend flow. The duct walls have a conductivity according to $c = 0.052$. The theoretical result obtained on the basis of an asymptotic analysis valid for $Ha, N \to \infty$ is indicated at the y-axis ($\rightarrow$). Experimental results for ($\circ$) $Ha = 7651$, ($\square$) $Ha = 3975$, (Δ) $Ha = 1992$.

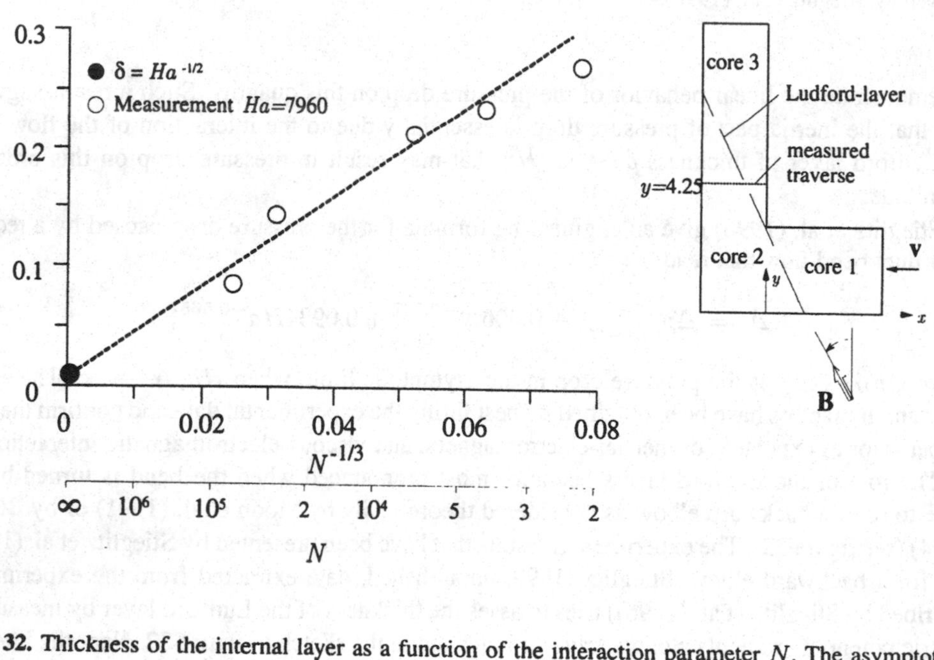

Figure 32. Thickness of the internal layer as a function of the interaction parameter N. The asymptotic theory valid for $Ha, N \to \infty$ gives a value as indicated on the axis. With increasing inertia the layers become thicker according to $\delta \sim N^{-1/3}$. The data has been evaluated by Stieglitz (1999, unpublished)

ations there exists a strong electrical flow coupling between neighboring channels. The coupling is much more pronounced as it would be for an array of straight ducts (Molokov (1993)). One has to expect unequal flow distribution among bends, if the same pressure difference drives the flows in each of them. The experimental results for multi-bend flows published by Stieglitz and Molokov (1997) show that the inertial part of pressure drop scales again with $N^{-1/3}$ and that for high values of N a good agreement with the asymptotic theory is achieved.

5.6 Fringing magnetic fields

In MHD-channel flow three dimensional flow may be induced either by changes in the geometry like in bends and expansions, or by a flow within a spatially varying magnetic field. This is the case when the flow enters or leaves the magnetic field. We will next discuss a three-dimensional flow redistribution in a straight channel induced by a spatially varying, i.e. fringing, external magnetic field. The situation is depicted in figure 33. Given an external magnetic field which decreases from a certain higher level of intensity to a lower level over some distance L. The induced voltage between the channel side walls is higher in the range of the higher intensity of B indicated by symbols ⊕⊕ and ⊖⊖ compared to the voltage in the lower intensity range labelled by ⊕ and ⊖. Thus a voltage difference along the range of the varying B-field is generated which drives a current in flow direction in one half section of the channel and in opposite direction in the second half section. These currents close across the channel width at upstream and downstream positions of the B-field transition zone as indicated in figure 33.

The local current density distribution with components in flow direction result in Lorentz forces with direction towards the side walls and corresponding transversal pressure gradient. Moreover, the upstream and downstream short circulated currents induce Lorentz forces opposite and in flow direction, which decelerate the core flow upstream and accelerate it downstream. As a consequence, the volumetric flux is expelled from the core flow region of the pipe to the vicinity of the side walls generating significant overspeed and an overall M-shape velocity profile. The redistribution of the volumetric flux gives rise to an additional "three-dimensional" pressure loss due to the flow redistribution. In figure 33d the velocity redistribution is demonstrated by measured velocity values in a circular duct at two positions (see figure 33d), at the top and the while it increases in side layer. Figure 33b demonstrates the pressure variation between the duct core and the side layer in the range of the 'fringing' B-field. The overall pressure gradient in flow direction is seen in figure 33c. A pronounced difference in pressure drop in comparison with the inertialess theory can not be identified. This indicates that inertia effects were not significant with regard to the overall pressure losses of inertialess MHD channel flow. Indeed, the comparison between the measured pressure loss and the calculated pressure losses shows a good agreement for this case.

Similar experiments have been performed by Barleon et al. (1989) for $c = 0.036$, $Ha = 7600$, $7.2 \cdot 10^3 < N < 2.5 \cdot 10^4$ which show the same quality of agreement between the asymptotic theory and the measured data.

Numerical simulations using the full set of MHD equations without simplifications have been performed e.g. by Sterl (1990), who finds the same behavior as outlined above already for relatively small Hartmann numbers such as $Ha = 50$. It has been shown by Lenhart and McCarthy (1991) that for Hartmann numbers like $Ha = 300$ MHD flow calculations in a fringing

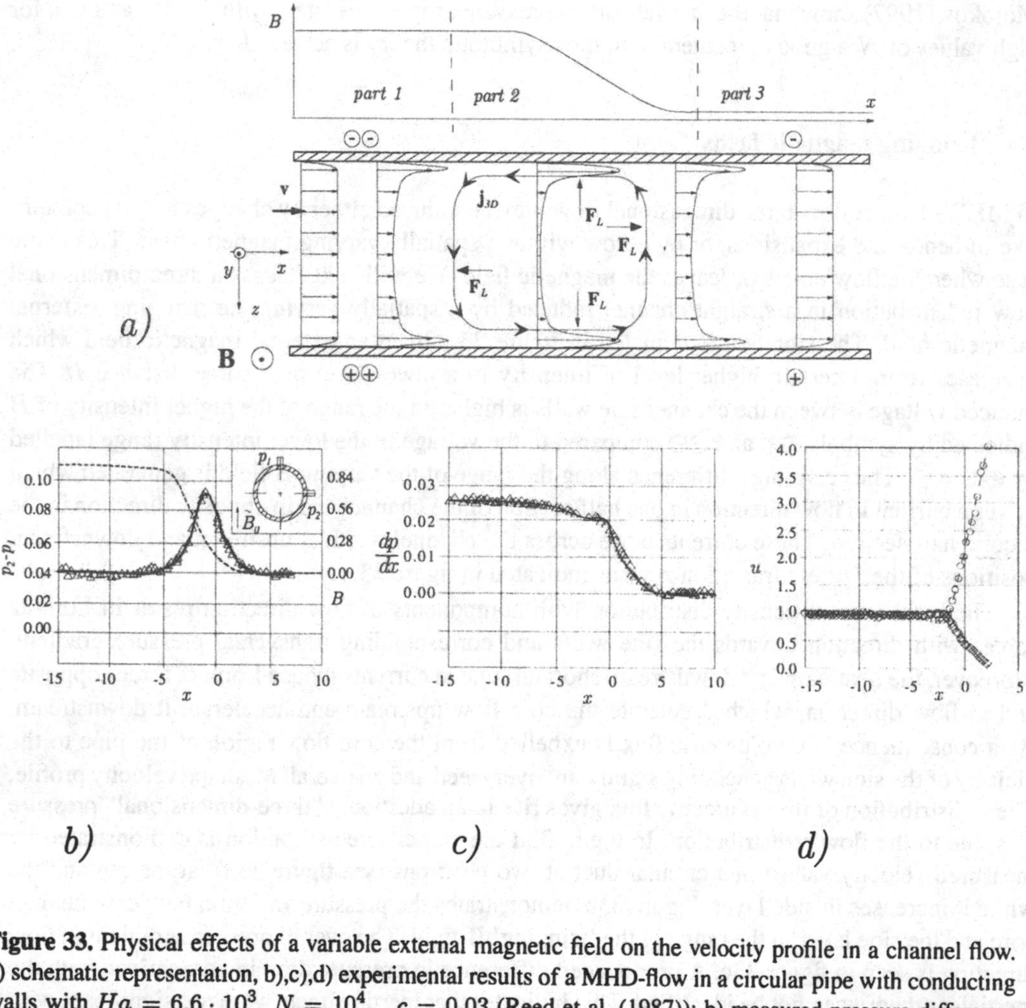

Figure 33. Physical effects of a variable external magnetic field on the velocity profile in a channel flow. a) schematic representation, b),c),d) experimental results of a MHD-flow in a circular pipe with conducting walls with $Ha = 6.6 \cdot 10^3$, $N = 10^4$, $c = 0.03$ (Reed et al. (1987)); b) measured pressure difference along the wall periphery versus the x-coordinate ($\triangle$), calculated values (—), normalized external magnetic field (- -); c) measured pressure gradient ($\triangle$) and calculated values (—); d) measured velocity distribution as a function of the x-coordinate at positions $z = 0.9$ ($\circ$) and $z = 0$ ($\diamond$), calculated distribution (—);

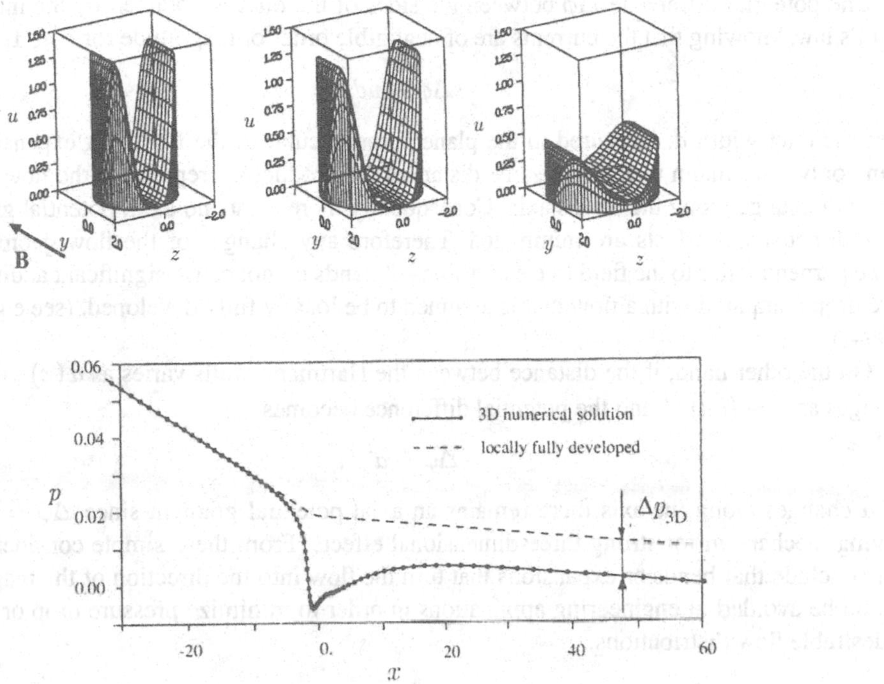

Figure 34. Three dimensional MHD flow in a circular expansion at $M = 1000$, $c = 0$. The duct has initially a radius of $r = 1$ and expands between $-2 < x < 2$ smoothly to $r = 2$. Velocity profiles at $x = 0$, $x = 1$, $x = 7$; pressure as a function of x compared with the value of a locally fully developed flow.

field performed on the basis of a complete numerical simulation and using asymptotic methods agree well.

5.7 Expansions

The MHD flow in expansions or contractions exhibits a strong similarity with the flows in a fringing magnetic field. While in the case of a fringing field the axial potential gradient is created by different values of the induced electric field $\mathbf{v} \times \mathbf{B}$ with varying values of $\mathbf{B}$, in the case of axially changing cross sections now the velocity $\mathbf{v}$ changes along the axis which may cause an axial potential difference that drives additional currents and creates thereby increased pressure drop. Additional currents create additional Lorentz forces and shape the flow profile toward an M-shape distribution see e.g. ? (?) or figure 34. Due to three-dimensional currents there remains an extra pressure drop Δp_{3D} in addition to that of an assumed locally fully developed flow.

The potential difference $\Delta\phi$ between the sides of the duct is obtained by the integration of Ohm's law, knowing that the currents are of negligible order of magnitude for $c \ll 1$. This yields

$$\Delta\phi \sim ud, \tag{166}$$

with the duct width d, measured in the plane perpendicular to the field. For expansions in this plane only (Hartmann walls have same distance) the product ud represents the flow rate which does not change along the duct's axis. Consequently there exists no axial potential gradient and three-dimensional effects are minimized. Therefore any changes of the flow geometry in the plane perpendicular to the field like expansions of bends do not cause significant additional pressure drop compared with a flow that is assumed to be locally fully developed. (see e.g. Molokov (1994))

On the other hand, if the distance between the Hartmann walls varies as $a(x)$, the velocity changes as $u \sim (ad)^{-1}$ and the potential difference becomes

$$\Delta\phi \sim a^{-1}. \tag{167}$$

As a changes along the axis there remains an axial potential gradient since $\Delta\phi = \Delta\phi(x)$ as driving mechanism for strong three-dimensional effects. From these simple considerations one can conclude that bends or expansions that turn the flow into the direction of the magnetic field should be avoided in engineering applications in order to minimize pressure drop or to exclude undesirable flow distributions.

5.8 Vortex generation

Vortex generation by shaping a unidirectional MHD flow such that internal shear layers become unstable has been considered in the past in a number of references (see e.g. Bühler (1996) and references therein). One example is outlined shortly.

The geometry used by Bühler (1996) is that of a Hartmann flow with insulating walls. Along the channel axis there are two strips of higher electrical conductance inserted into the walls as shown in figure 26. This leads to a reduction of velocity in the region of the strips as predicted by equation (55), where

$$u = \frac{c+1}{cHa+1} = \begin{cases} 1 \\ Ha^{-1} \end{cases} \text{for} \quad \begin{matrix} c = 0 \\ c \to \infty \end{matrix}. \tag{168}$$

Here, the velocity scale is used such that the velocity in the insulating part of the duct becomes unity. It is obvious that in the regions where the conductivity changes from insulating to conducting conditions the core velocity jumps. By this two shear layers are formed that loose their laminar stability once the Reynolds number exceeds a critical value. Beyond that threshold the flow becomes unstable and exhibits time-dependent behavior comparable to a Kármán vortex street. It can be shown by a detailed stability analysis that the critical Reynolds number depends linearly on the Hartmann number

$$Re_c \sim Ha, \tag{169}$$

for strong enough magnetic fields, for $cHa \gg 1$. The Threshold for instability has been confirmed by experiments performed by Debray (1997) and by Frank et al. (1997). A snapshot of the time-dependent motion can be seen in figure 36.

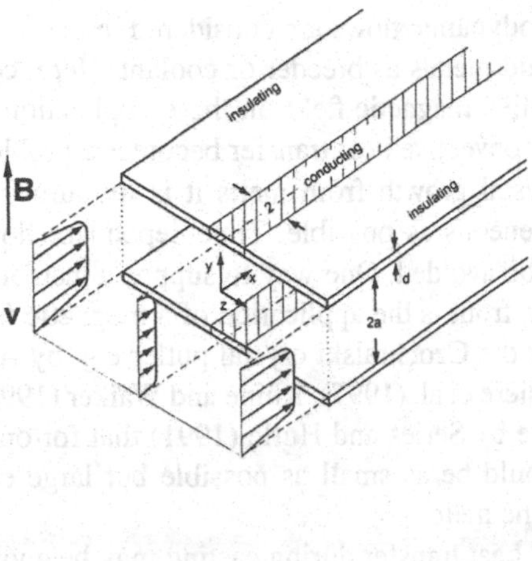

Figure 35. MHD flow in a duct with non-uniform conductance of the Hartmann walls. The characteristic scale is chosen as the half width of the conducting strip, that is aligned with the mean flow direction (see Bühler (1996)).

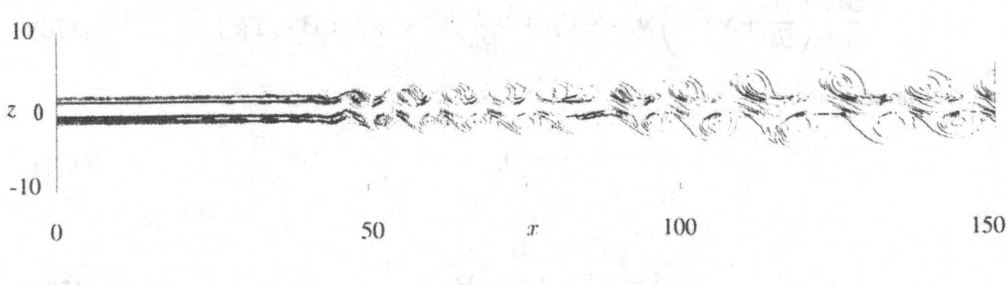

Figure 36. MHD vortex street. Isolines of vorticity for $Ha = 1000$, $Re = 500$ according to Bühler (1996)

6 Buoyancy driven MHD flows

Buoyancy driven magnetohydrodynamic flows are considered in the frame of magnetically confined fusion reactors using liquid metals as breeder or coolant. Here, convective or forced flow is strongly opposed by the applied magnetic field. In these applications it may happen that the MHD damping is so large that convective heat transfer becomes a problem.

During semi conductor crystal growth from melts it is the aim to have at the fluid-solid interface conditions as homogeneous as possible. Time-dependent flows as they might occur by buoyant convection should be avoided. One way to suppress instabilities or to minimize the velocities near the solidification front is the application of a magnetic field as it was studied for Bridgman crystal growth or for the Czochralski crystal puller e.g. by Alboussière et al. (1995), Ma and Walker (1996), Alboussière et al. (1997), Khine and Walker (1998), Walker (1998). It has been reported in a review article by Series and Hurle (1991) that for optimal growth conditions the magnetic field strength should be as small as possible but large enough to suppress time dependent turbulent motion in the melt.

To control the conditions of heat transfer during casting may be another application of magnetic fields on buoyant flows.

During this section the attention is focussed on the solution of simple examples of magneto-convective flows with the aim to get some insight into the phenomena and the typical scales for the variables. Additionally some results for more complex applications are outlined and some experimental results will be shown.

6.1 Governing equations

The inductionless buoyant flow of an incompressible viscous electrically conducting fluid is governed by the equation of motion

$$\frac{Gr}{Ha^4}\left(\frac{\partial}{\partial t}+\mathbf{v}\cdot\nabla\right)\mathbf{v}=-\nabla p+\frac{1}{Ha^2}\nabla^2\mathbf{v}+\mathbf{j}\times\mathbf{B}-T\hat{\mathbf{g}}\,, \tag{170}$$

the conservation of mass

$$\nabla\cdot\mathbf{v}=0, \tag{171}$$

Ohm's law

$$\mathbf{j}=-\nabla\phi+\mathbf{v}\times\mathbf{B}, \tag{172}$$

and conservation of charge

$$\nabla\cdot\mathbf{j}=0. \tag{173}$$

Equations similar to those displayed above have been used e.g. in Ma and Walker (1995), Ma and Walker (1996) for calculations of magneto-convection during Czochralski crystal growth applications, while e.g. Albousière et al. (1993), Alboussière et al. (1996), Mößner (1996), Ben Hadid and Henri (1997) prefer scales different from these.

Here, $\mathbf{B}$, $\mathbf{v} = (u, v, w)$, $\mathbf{j} = (j_x, j_y, j_z)$, and ϕ denote the magnetic field, velocity, electric current density, and the electric potential, scaled by the reference quantities B_0, $v_0 = \nu/L\ Gr/Ha^2$, $j_0 = \sigma v_0 B_0$, $\phi_0 = L v_0 B_0$. The unit vector in the direction of gravity is $\hat{\mathbf{g}}$. The variable T represents the difference between a the local temperature and a reference temperature T_0, scaled by a characteristic temperature difference $\Delta T = \alpha L$. The variable α is the magnitude of a temperature gradient typical for the problem under consideration. The density of the fluid at temperature T_0 is ρ_0, the thermal expansion coefficient according to the *Boussinesq approximation* is β. The difference between the local pressure and the isothermal hydrostatic pressure (at T_0) scaled by $L j_0 B$ is called p. The electric conductivity of the fluid is σ and L stands for a typical length scale measured in the direction of the magnetic field. The kinematic viscosity is denoted by ν.

The non-dimensional parameters are the

$$\textit{Hartmann number } Ha = L B_0 \sqrt{\sigma/\rho_0 \nu} \tag{174}$$

and the

$$\textit{Grashof number } Gr = \beta g \alpha L^4/\nu^2. \tag{175}$$

The square of the Hartmann number gives the ratio of electromagnetic to viscous forces. The Grashof number quantifies the importance of buoyant effects. The velocity scale is given by the viscous scale ν/L times Gr/Ha^2, where the ratio Gr/Ha^2 corresponds to a Reynolds number. The ratio Ha^4/Gr is the square of the Lykoudis number and corresponds to an interaction parameter.

The scales introduced above are especially useful for strong magnetic fields. The buoyancy forces and Lorentz forces are treated on the same order of magnitude while viscous terms become less important in most of the fluid region, except within very thin boundary layers. Inertial terms seem to play no significant role unless the Grashof number becomes of the order Ha^4 or larger.

The temperature distribution is governed by an energy balance

$$Pe\left(\frac{\partial}{\partial t} + \mathbf{v}\cdot\nabla\right) T = \nabla^2 T + Q. \tag{176}$$

The quantity Q is the volumetric heat source scaled by $\lambda \Delta T/L^2$, caused e.g. by viscous or Ohmic dissipation or due to nuclear irradiation. The thermal conductivity of the fluid is λ. The

$$\textit{Peclet number } Pe = v_0 L/\kappa$$

gives the ratio of convective to conductive heat flux. The thermal diffusivity is denoted by κ. The Peclet number reads in quantities already introduced above as

$$Pe = \frac{Gr}{Ha^2} Pr, \tag{177}$$

where Pr is the

$$\textit{Prandtl number } Pr = \nu/\kappa. \tag{178}$$

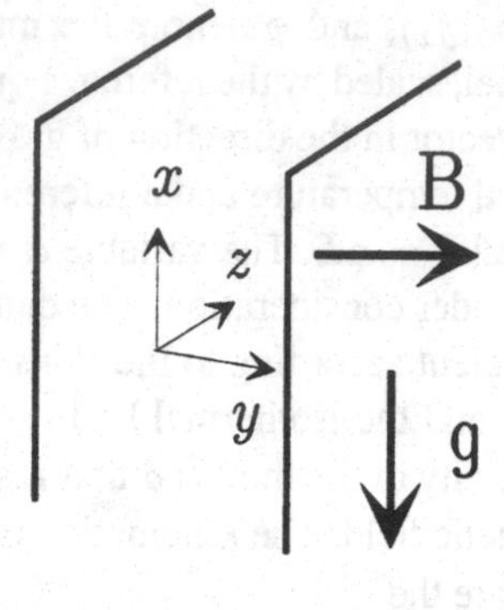

Figure 37. Differentially heated vertical gap. Geometry and coordinates

The boundary conditions are the no-slip condition at the duct walls

$$\mathbf{v} = 0,$$

and the *thin wall condition* for electric currents

$$\mathbf{j} \cdot \mathbf{n} = c\nabla_w^2 \phi. \tag{179}$$

The subscript ' $_w$ ' denotes properties at the wall and ∇_w^2 is the two-dimensional Laplacian in the plane of the wall. The constant c is called the

$$\textit{wall conductance ratio } c = \sigma_w t \; .$$

The ratio of the electric wall to fluid conductivity is σ_w. The non-dimensional thickness of the wall t is assumed to be small, $t \ll 1$. Equation (179) ensures charge conservation in the plane of the wall; currents leaving the fluid enter the wall balance as a source term and create inside the wall a potential distribution. It is further assumed that there is no contact resistance at the fluid wall interface so that the fluid potential ϕ at the wall is equal to the wall potential. Note, $\mathbf{n}$ is the *inward* unit normal to the duct wall.

The thermal conditions are

$$T = T_w \quad \text{or } \mathbf{n} \cdot \nabla T = -q_w$$

for perfectly thermally conducting walls or for a given non-dimensional wall heat flux q_w.

6.2 Differentially heated vertical gap

As a first example the steady state magnetoconvective flow in a layer between two differentially heated vertical walls is analyzed. For this case there exists an analytic solution that is valid also for flows in closed cavities in some distance from the other walls that are parallel to the magnetic field.

Here, the walls are located at the positions $y = \pm 1$, and kept isothermal at temperatures $T(\pm 1) = \pm 1$. The uniform magnetic field is perpendicular to the walls $\mathbf{B} = \hat{\mathbf{y}}$ and gravity acts in negative x-direction, $\hat{\mathbf{g}} = -\hat{\mathbf{x}}$. If the plates extend far enough along the vertical direction a fully developed flow will establish in the region considered here, in some distance from their ends. The pressure gradient becomes constant

$$\nabla p = k\hat{\mathbf{x}}, \tag{180}$$

and other variables do not change along x. For such conditions the velocity $\mathbf{v} = u\hat{\mathbf{x}}$ is unidirectional. It is assumed that the extension of the plates along the horizontal z-direction is large that changes $\partial/\partial z$ can be neglected. This holds even for the electric potential because there is no net flux of current along z. For this geometry the potential becomes uniform and plays no longer a role in the analysis. The magnitude k of the pressure gradient is determined by the net flux of fluid in the vertical direction. For closed cavities $\int u dy = 0$.

The velocity and temperature then depend only on the wall normal (the magnetic field) direction, $u = u(y)$, $T = T(y)$. Moreover, the energy equation decouples from the velocity field and, when volumetric heating is absent, the temperature is found as

$$T = y. \tag{181}$$

The momentum balance simplifies with Ohm's law, $j_z = u$, to

$$\frac{1}{Ha^2}\frac{\partial^2 u}{\partial y^2} - u + T = k. \tag{182}$$

The solution that satisfies no slip at the walls, $u(\pm 1) = 0$, reads

$$u = y - \frac{\sinh(Hay)}{\sinh(Ha)}. \tag{183}$$

Velocity profiles for several values of the Hartmann number are shown in figure 38. The solution has a linear dependence of the core velocity along y and exhibits the typical exponential Hartmann boundary layers at the walls, if the Hartmann number is large. The role of viscosity in the Hartmann layer is to match smoothly the core solution with the no slip condition at the wall.

For large Hartmann numbers one finds the asymptotic relation

$$u \approx y \mp \mathrm{e}^{Ha(|y|-1)} \quad \text{for } Ha \gg 1, \quad y \gtrless 0. \tag{184}$$

For small Hartmann numbers the hydrodynamic limit is approached as

$$\frac{u}{Ha^2} \approx \frac{1}{6} y\left(1 - y^2\right) \quad \text{for } Ha \ll 1. \tag{185}$$

A nondimensional quantity of vertically convected heat is (Blūms et al. (1987))

$$q = Pe \int_{-1}^{1} uT dy \tag{186}$$

which evaluates to

$$\frac{1}{2Pe} q = \frac{1}{3} - \frac{1}{Ha}\coth(Ha) + \frac{1}{Ha^2}. \tag{187}$$

Recall that the Peclet number scales as $Pe \sim Gr/Ha^2$; it become obvious that the convective heat transfer is negligible for high Hartmann numbers.

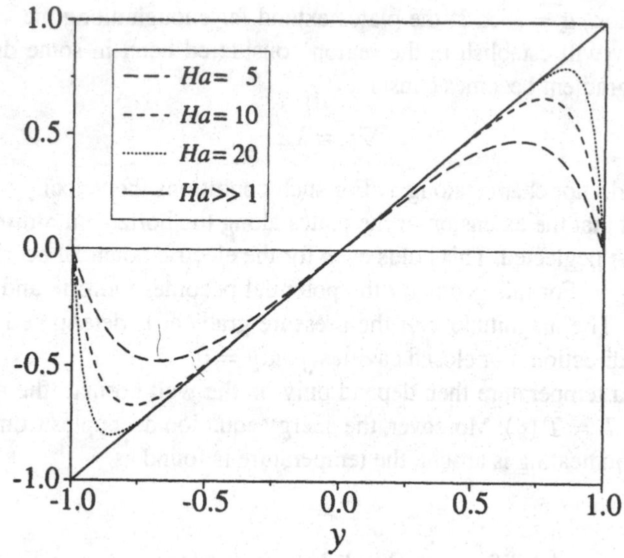

Figure 38. Velocity in a differentially heated gap with $T = y$.

6.3 Differentially heated vertical channel

Any geometry in engineering applications has finite dimensions. To approach real situations assume that the gap is now bounded in z-direction by two additional adiabatic walls, say at $z = \pm b$. At the new walls the boundary conditions are $u = \partial T/\partial z = 0$. With these walls the problem now depends on the z-coordinate with the consequence that lateral potential gradients enter the analysis. By symmetry the solution can be restricted to one quarter of the duct with $\phi = u = 0$ and $\phi = \partial u/\partial z = 0$ at $y = 0$ and $z = 0$, respectively.

By differentiation with respect to z of the equation (170) and by assuming fully developed flow along the channel axis one finds

$$\frac{1}{Ha^2}\nabla^2\frac{\partial u}{\partial z} + \frac{\partial j_y}{\partial y} = -\frac{\partial T}{\partial z}, \tag{188}$$

when the charge conservation equation (173) has been used to eliminate $\partial j_z/\partial z$. With the y component of Ohm's law one finds at leading order of approximation (in the core)

$$\frac{\partial j_y}{\partial y} = -\frac{\partial T}{\partial z}. \tag{189}$$

an integration along y yields with $j_y(y = 0) = -\phi_H$

$$j_y(y) = -\frac{\partial}{\partial z}\int_0^y T dy - \phi_H\,, \tag{190}$$

and when introduced into the thin-wall condition (179) one arrives at

$$c_H \frac{\partial^2 \phi_H}{\partial z^2} = \frac{\partial \bar{T}}{\partial z} + \phi_H \quad \text{at } y = 1 \tag{191}$$

for the potential at the Hartmann wall, where $\bar{T}$ abbreviates the y-averaged temperature. The variable ϕ_H stands for the core potential at the Hartmann wall and it equals the potential inside the wall if it is assumed that the potential does not change across the Hartmann layer to the leading order of approximation (see e.g. Moreau (1990)). The equation (191) holds under the assumption that the Hartmann walls are much better conducting than the Hartmann layer, when $c \gg Ha^{-1}$. If the conductivity of the side walls is high enough, $c \gg Ha^{-1/2}$, the core currents enter the side walls unchanged and create there a potential distribution according

$$c_s \frac{\partial^2 \phi_s}{\partial y^2} = -T + k \quad \text{at } z = b. \tag{192}$$

At the corner ($y = 1, z = b$) the wall potentials are continuous $\phi_H = \phi_s$ and charge conservation at the corner requires $c_H \frac{\partial \phi_H}{\partial z} = -c_s \frac{\partial \phi_s}{\partial y}$. After the wall potentials are known one can integrate equation (190) once more to obtain the potential at any point inside the core, by which the core velocity

$$u_c = -\frac{\partial \phi}{\partial z} - T + k \tag{193}$$

is finally determined. One can show that the solution in the Hartmann layers keeps the exponential profile as already discussed in the previous subsection. The composite core-Hartmann layer solution then becomes

$$u(y, z) = u_c(y, z) - u_c(1, z)\left(1 - e^{Ha(y-1)}\right). \tag{194}$$

The solution in the side layers, near the walls which are parallel to the magnetic field can be obtained by using a stretched coordinate $\zeta = Ha^{1/2}(b - z)$. For large Hartmann numbers the equation determining the viscous correction φ to the potential then simplifies to

$$\frac{\partial^4 \varphi}{\partial \zeta^4} = \frac{\partial^2 \varphi}{\partial y^2}, \tag{195}$$

which can be solved by separation of variables. One finds a general solution in the form

$$\varphi = \sum \left[A_i \cos(\alpha_i \zeta) + B_i \sin(\alpha_i \zeta)\right] e^{-\alpha_i \zeta} \begin{Bmatrix} \cos(\beta_i y) \\ \sin(\beta_i y) \end{Bmatrix} \quad \text{for } i = \begin{Bmatrix} 1, 3, 5, \dots \\ 2, 4, 6, \dots \end{Bmatrix}, \tag{196}$$

where $\beta_i = \frac{1}{2}\pi i$ and $\alpha_i = \sqrt{\beta_i/2}$. The remaining coefficients A_i and B_i are determined by a matching of the combined core-side layer solution with the solution at the side wall for velocity and potential (see e.g. Bühler (1998)). The method of solution in the side layer is essentially the same as for pressure driven duct flows except that the right hand side of equation (192) now depends on y.

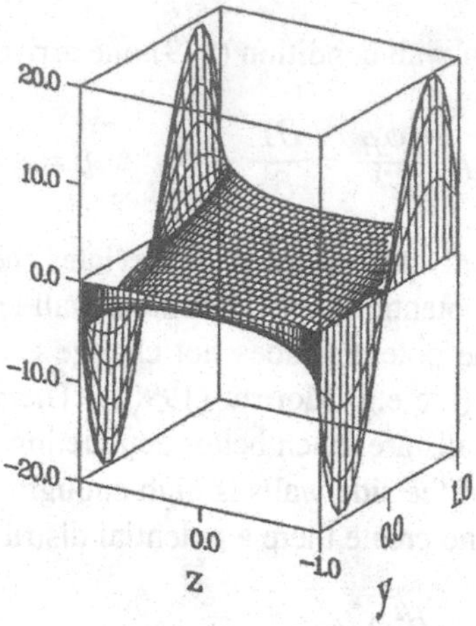

Figure 39. Velocity for buoyancy driven flows with $T = y$, $M = 1000$, $c_H = 0.05$, $c_s = 0.1$ (Bühler (1998))

A typical velocity profile is shown in figure 39. One observes that the core velocity is linear along magnetic field lines with thin Hartmann layers. Approaching the sides the magnitude of velocity increases in the core. This is due to the fact that equation (191) crates solutions with exponential layers near the sides with thickness $\delta \sim c_H^{1/2}$ for small c. Closer to the side wall one observes the viscous side layers with thickness $\delta \sim Ha^{-1/2}$ and magnitude of velocity on the order $Ha^{1/2}$.

6.4 Horizontal layer with horizontal temperature gradient

This subsection deals with the buoyancy driven flow in a horizontal gap with axial temperature gradient. A geometry of this type is related to applications in horizontal Bridgman crystal growth. The liquid solidifies at the cold side from a melt that has approximately a uniform axial temperature gradient $\nabla T \approx -\hat{\mathbf{x}}$ in some distance from the solidification front. The application of a magnetic field may help to suppress or control the buoyant flow during crystal growth for an improvement of the quality of the solidified material.

For the case when a magnetic field is perpendicular to the horizontal plates Garandet et al. (1992) propose a two-dimensional solution to the problem. They consider first a unidirectional flow $\mathbf{v} = (u, 0, 0)$ that is valid if the fluid region extends long enough (see figure 40). Later they study the turning flow near the solid crystal.

The solution for fully developed flow between the two confining plates evaluates to just the same as for the flow in the vertical gap discussed above in subsection (6.2). The reason for a coincidence of both solutions with different directions of heat flux and different orientations of gravity (here, $\hat{\mathbf{g}} = (0, -1, 0)$) is not obvious at a first view. However, by taking the curl of the momentum equation both cases lead to exactly the same equation for the vorticity $\omega = \nabla \times (u\hat{\mathbf{x}})$, a fact that allows to use also the results obtained for a differentially heated vertical channel

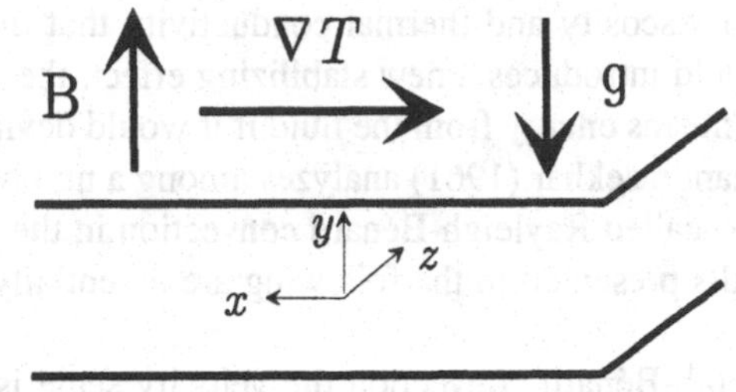

Figure 40. Geometry typical for horizontal Bridgman crystal growth

(subsection 6.3) as solution for the flow in a horizontal duct with finite cross section in an axial temperature gradient. The vorticity equation becomes

$$\frac{1}{Ha^2}\nabla^2\omega + \frac{\partial}{\partial y}\mathbf{j} + \hat{\mathbf{z}} = 0, \tag{197}$$

with the buoyant term $\nabla \times (\hat{\mathbf{g}}T) = -\hat{\mathbf{g}} \times \nabla T = \hat{\mathbf{z}}$. The z component is identical to the equation (182), differentiated with respect to y, and the y component becomes equal to equation (188) if in the former case the temperature gradient was $\nabla T = \hat{\mathbf{y}}$.

The results described in the previous subsection are in accordance with Albousiére et al. (1993) who analyze among a number of cross sections the flow in a horizontal rectangular shape duct with axial temperature gradient and agree qualitatively with those obtained by Ben Hadid and Henri (1997) although the latter authors use different electrical boundary conditions. It seems to be a common feature of buoyant MHD flow in rectangular cavities that a large fraction of the flow is carried near the side walls in high velocity jets, while the velocities in the core are much lower.

The flow in the horizontal layer modifies the temperature field by terms on the order of Pe. The modified temperature will not affect the solution for the flow since only the unchanged horizontal temperature gradient enters the analysis.

Near the solidification front the flow turns in a layer of thickness on the order $Ha^{-1/2}$ with magnitude of velocity proportional to $Ha^{1/2}$. Garandet et al. (1992) conclude that inertia effects are almost unimportant for $Gr/Ha^3 \ll 1$, a condition that holds usually for horizontal Bridgman growth.

6.5 Horizontal layer with vertical temperature gradient

Consider next a horizontal layer of an electrically conducting viscous incompressible fluid as sketched in figure 40, but assume now that all the three quantities are parallel, the magnetic field $\mathbf{B} = \hat{\mathbf{y}}$, gravity $\hat{\mathbf{g}} = -\hat{\mathbf{y}}$, and the applied temperature gradient $\nabla T = -\hat{\mathbf{y}}$. It is assumed that the isothermal walls are located at the positions $y = \pm\frac{1}{2}$ with temperatures $T = \mp\frac{1}{2}$. The hotter wall is at the bottom, the colder at the top. The governing equations are scaled for the present problem with the gap width L and all relevant nondimensional groups are built with this reference length.

The Governing equations (170-173) allow for a motionless steady state solution with $\mathbf{v} = \mathbf{j} = 0$, $T = -y$. The fluid layer with unstable thermal stratification is maintained at this steady state by the stabilizing effects due to viscosity and thermal conductivity that dissipate small perturbations. An applied magnetic field introduces a new stabilizing effect, the Joule dissipation which removes by electromagnetic means energy from the fluid if it would deviate from the motionless state. In his famous work Chandrasekhar (1961) analyzes among a number of stability problems the present situation of the so-called Rayleigh-Bénard convection in the presence of a magnetic field and the theoretical results presented in the following are essentially those presented in his text book.

For an analysis of Rayleigh-Bénard convection the velocity scale is used as $v_0 = \kappa/L$, a velocity that accounts for the speed at which thermal disturbances may propagate in a quiescent fluid. Using this scale the governing equations take the form

$$\frac{1}{Pr}\left(\frac{\partial}{\partial t} + \mathbf{v}\cdot\nabla\right)\mathbf{v} = -\nabla p + \nabla^2\mathbf{v} + Q\mathbf{j}\times\mathbf{B} - RaT\hat{\mathbf{g}}\,, \tag{198}$$

and

$$\left(\frac{\partial}{\partial t} + \mathbf{v}\cdot\nabla\right)T = \nabla^2 T \tag{199}$$

for conservation of momentum end energy, respectively. The other equations remain unchanged. These equations contain the dimensionless parameters, the

$$\textit{Chandrasekhar number } Q = Ha^2 = \frac{L^2 B_0^2 \sigma}{\rho_0 \nu} \tag{200}$$

and the

$$\textit{Rayleigh number } Ra = GrPr = \frac{\beta g \Delta T L^3}{\nu\kappa}. \tag{201}$$

Chandrasekhar (1961) showed that under certain circumstances a steady motion is initiated once the temperature difference between the plates, say the Rayleigh number, exceeds a critical value Ra_c. Near marginal stability (critical conditions) the flow will be weak so that all magnitudes for $\mathbf{v}$ and $\mathbf{j}$ are small and the temperature varies with the small perturbation θ as $T = -y+\theta$. The boundary conditions at rigid isothermal walls are $\mathbf{v} = \theta = 0$. It turns out during the analysis that the currents near the walls are parallel to the walls so that the wall conductivity does not enter the problem and no electric boundary conditions are required.

Once a convective regime is established one can observe an increased heat flux q. In addition to the conductive one $q_0 = -\lambda\Delta T/L$ there is now a heat transport due to the fluid motion. A nondimensional parameter that characterizes this quantity is the

$$\textit{Nusselt number } Nu = \frac{q}{q_0}. \tag{202}$$

Convective heat transfer exists for $Nu > 1$, while $Nu = 1$ describes the pure conductive state.

Without going into the details of Chandrasekhar's analysis basic ideas are outlined and some results are shown below. By assuming spatially periodic solutions with a horizontal wave number a and assuming the onset of motion to be steady state one can eliminate variables and arrives at a 6th order ordinary, homogeneous differential equation that states an eigenvalue problem in combination with the corresponding boundary conditions. This equation reads for the vertical component of velocity

$$(D^2 - a^2)\left[(D^2 - a^2)^2 - QD^2\right] W + Ra^2 W = 0, \tag{203}$$

$$W = 0,\ DW = 0,\ \left[(D^2 - a^2)^2 - QD^2\right] W = 0,\ \text{ for } y = 0 \text{ and } 1. \tag{204}$$

For the case of free slip conditions at a wall the relation $DW = 0$ has to be replaced by $D^2W = 0$. In these equations W stands for the variation of vertical velocity along the vertical direction and D is used as abbreviation for the operator d/dy. The equation admits non-trivial solutions only for certain combinations of the control parameters Ra and Q with the wave number a. Chandrasekhar deduces a relation for the critical Rayleigh number at the onset of fluid motion of the form

$$Ra_c = \frac{\pi^2 + a^2}{a^2}\left[(\pi^2 + a^2) + \pi^2 Q\right], \tag{205}$$

where the critical wave number is obtained as a root of the polynomial

$$2x^3 + 3x^2 = 1 + \frac{Q}{\pi^2}, \quad \text{where } x = \left(\frac{a}{\pi}\right)^2. \tag{206}$$

This analytical result has been obtained for free slip boundary conditions at the plates. Results which account for no slip have been obtained by Chandrasekhar as well but they can not be displayed in a closed form. Both, the free-slip and the no-slip results are plotted in the following figures 41, 42.

One finds the limiting hydrodynamic case when $Q \to 0$ that $Ra_c \to \frac{27}{4}\pi^4$, (1707.76) and $a_c \to \frac{\pi}{\sqrt{2}}$, (3.117) for free slip (no slip) walls. On the other hand Ra_c and a_c approach the asymptotic relations

$$Ra_c \to \pi^2 Q, \quad a_c \to \left(\frac{\pi^4}{2} Q\right)^{1/6} \quad \text{as } Q \gg 1. \tag{207}$$

The onset of convection is strongly delayed by a magnetic field.

The theoretical results agree well with known experiments as reported by Chandrasekhar (1961). In addition to the experimental data published by Chandrasekhar recent experimental results published by Burr et al. (1999b) are added to the figure 41. The critical Rayleigh number has been evaluated from the latter reference as the point where the convective heat transfer sets in, where the Nusselt number rises above unity.

The critical wave number increases, the wave length $\lambda = 2\pi/a$ decreases with the intensity of the magnetic field. The convection patterns decrease their lateral dimensions in order to shorten the flow path perpendicular to the magnetic field and to minimize Joule dissipation. On the other hand, viscous effects (and thermal conductance) may terminate fluid motion if the lateral dimensions become too small. Generally, the patterns depend on Q and *"become narrow and elongated"* with increasing Q.

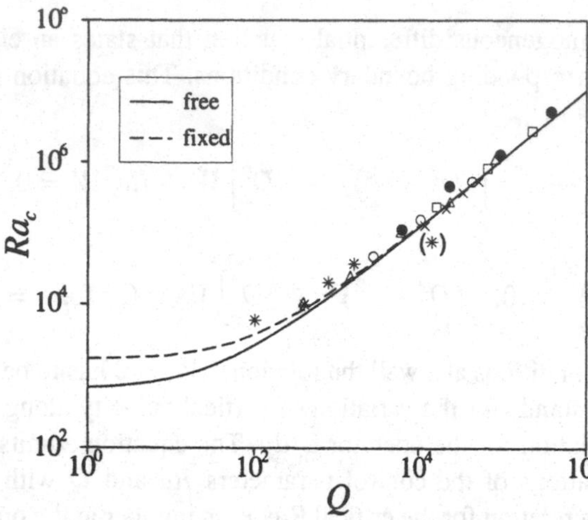

Figure 41. Critical Rayleigh number for the onset of Rayleigh-Bénard convection confined between no slip and free slip plates. Comparison with experiments as shown by Chandrasekhar (1961). Results from recent experiments (Burr et al. (1999b)) marked as "*" are added.

It is worth to notice that in the asymptotic regime as $Q \to \infty$ the instability sets in for a fixed temperature difference and viscosity disappears from the stability condition. Then, the quantity $L^2 B_0^2 \sigma / \rho_0 = \nu_{eff}$ may take the role of an effective viscosity. As $Q \to \infty$ one finds that $a_c \to 0$ and *"it is viscosity that prevents the cells from collapsing into lines"*.

Note, the equations (198, 199) displayed above are valid in the inductionless limit, when the influence of the motion on the magnetic field is negligible. If this is not guarantied one can find with a more general analysis using in addition a transport equation for the induced magnetic field, an onset of convection as time-dependent motion (overstability). The results then depend on both, the thermal Prandtl number Pr and the magnetic Prandtl number P_m. For $P_m < Pr$, a case that is always met in liquid metal MHD, the onset of convection is stationary.

After a fluid motion sets in the heat transfer is increased. Results for the additionally transported heat measured in terms of Nusselt numbers are plotted in figure 43. Here, the results are plotted versus the ratio Ra/Q, which should tend at marginal stability for $Q \to \infty$ to the unique value $Ra/Q \to B = \pi^2$, with $Nu = 1$. The experiment detects a measurable increase of heat flux, $Nu > 1$ at values Ra/Q that are larger than the predictions; here, $B = 28$. If one takes into account that some weak convective flow may exist even before the agreement is fairly good. One can extract a heat transfer correlation of the form

$$\frac{Nu - 1}{Q^{1/2}} = A\left(\frac{Ra}{Q} - B\right)^{2/3} \text{ with } A = 1.1 \cdot 10^{-3}, \quad B = 28 \tag{208}$$

that fits all measured data with reasonable accuracy, especially those at large Q.

Burr et al. (1999b) find at higher supercritical conditions a bifurcation from steady state to time-dependent flow regimes. They identify the threshold for this bifurcation and analyze in detail the time history of experimental signals.

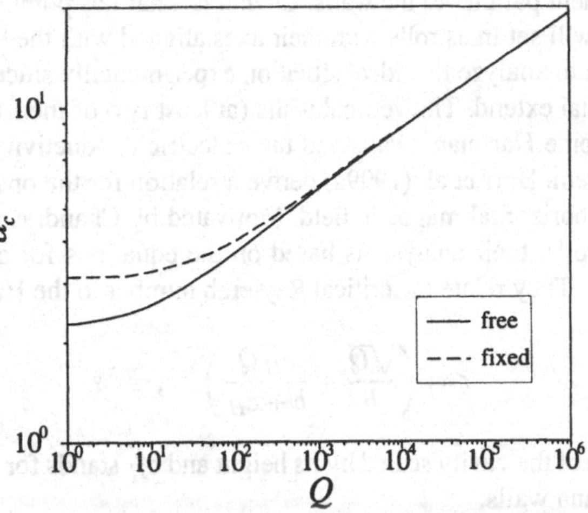

Figure 42. Critical wave number for the onset of Rayleigh-Bénard convection confined between rigid and free plates.

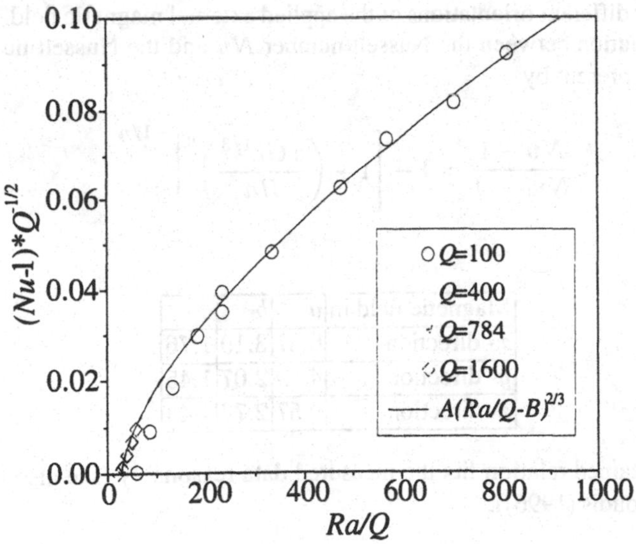

Figure 43. Nusselt numbers for MHD Rayleigh-Bénard convection. Data taken from Burr et al. (1999b).

One can imagine other situations of Rayleigh-Bénard convection, when for example the magnetic field has a component parallel to the walls. Chandrasekhar has pointed out that under such circumstances the flow will set in as rolls with their axes aligned with the horizontal component of the field. It is difficult to analyze this ideal situation experimentally since any experiment will be limited in its horizontal extend. The vertical walls (at least two of them) that close the experimental cavity then become Hartmann walls and their electric conductivity now is essential for the solution of the problem. Burr et al. (1999a) derive a relation for the onset of convective flow in a finite cavity with a horizontal magnetic field. Motivated by Chandrasekhar's statement that the flow sets in as 2D rolls, their analysis is based on the equations for quasi 2D MHD flows shown by Bühler (1996). They relate the critical Rayleigh number to the Hartmann braking time

$$\tau = \left(\frac{\sqrt{Q}}{b} + \frac{c_H Q}{b + c_H}\right)^{-1}, \tag{209}$$

where b is the half width of the cavity scaled by its height and c_H stands for the wall conductance parameter of the Hartmann walls.

For strong magnetic fields the following asymptotic relation is obtained

$$Ra_c = 4\pi\tau^{-1}, \quad a_c = \pi, \text{ for } Ha \to \infty \tag{210}$$

and confirmed by the experiment. Furthermore the problem is analyzed in detail with respect to heat transfer and time series of signals. A comprehensive description would be out of the scope of the present review.

The buoyant heat transfer in cubic cavities has been investigated experimentally by Okada and Ozoe (1992) for different orientations of the applied external magnetic field. They summarize their results by a relation between the Nusselt number Nu and the Nusselt number Nu_0 when no magnetic field is present by

$$\frac{Nu - 1}{Nu_0 - 1} = 1 - \left[1 + \left(\frac{a\, Gr^{1/3}}{Ha}\right)^b\right]^{-1/n}, \tag{211}$$

with

Magnetic field in	a	b	c
x- direction	0.57	3.19	1.76
y- direction	4.19	2.07	1.45
z- direction	0.57	2.72	1.44

This empirically obtained relation fits the measured data reasonably (see also the discussion of this formula by Lykoudis (1996)).

References

Albousiére, T., Garandet, J., and Moreau, R. (1993). Buoyancy-driven convection with a uniform magnetic field. part 1. asymptotic analysis. *Journal of Fluid Mechanics* 253:545–563.

Alboussière, T., Neubrand, A. C., Garandet, J. P., and Moreau, R. (1995). Magnetic field and segregation duringBridgeman growth. *Magnetohydrodynamics* 31:228–235.

Alboussière, T., Garandet, J. P., and Moreau, R. (1996). Asymptotic analysis and symmetry in MHD convection. *Physics of Fluids* 8(8):2215–2226.

Alboussière, T., Neubrand, A. C., Garandet, J. P., and Moreau, R. (1997). Segregation during horizontal Bridgeman growth under an axial magnetic field. *Journal of Crystal Growth* 181:133–144.

Alty, C. J. N. (1971). Magnetohydrodynamic duct flow in a uniform transverse magnetic field of arbitrary orientation. *Journal of Fluid Mechanics* 48:429–461.

Barleon, L., Lenhart, L., Mack, H. J., Sterl, A., and Thomauske, K. (1989). Investigations on liquid metal MHD in straight ducts at high Hartmann numbers and interaction parameters. In Müller, U., Rehme, K., and Rust, K., eds., *Proceedings of the 4th International Topical Meeting on Nuclear Reactor Thermal-Hydraulics, Karlsruhe, October 10-13*. G. Braun, Karlsruhe. 857–862.

Barleon, L., Mack, K. J., Kirchner, R., Frank, M., and Stieglitz, R. (1995). MHD heat transfer and pressure drop in electrically insulated channels at fusion relevant parameters. In Herschbach, K., Maurer, W., and Vetter, J. E., eds., *Fusion Technology 1994*. Elsevier. 1201–1204.

Ben Hadid, H., and Henri, D. (1997). Numerical study of convection in a horizontal Bridgeman configuration under the action of a constant magnetic field. Part 2. three-dimensional flow. *Journal of Fluid Mechanics* 333:57–83.

Blūms, E., Mikhailov, Y., and Ozols, R. (1987). *Heat and mass transfer in MHD flows*. World Scientific Publishing Co. Pte. Ltd.

Branover, H., Vasil'ev, H., and Gelfgat, Y. (1967). Hydraulic resistance of MHD pipes. *Magnitnaya Gidrodynamica* 4(3):1.

Branover, H. (1978). *Magnetohydrodynamic flow in ducts*. John Wiley & Sons, New York, Toronto.

Bühler, L. (1994). Magnetohydrodynamic flows in arbitrary geometries in strong, nonuniform magnetic fields.-a numerical code for the design of fusion reactor blankets. *Fusion Technology* 27:3–24.

Bühler, L. (1996). Instabilities in quasi-two-dimensional magnetohydrodynamic flows. *Journal of Fluid Mechanics* 326:125–150.

Bühler, L. (1998). Laminar buoyant magnetohydrodynamic flow in vertical rectangular ducts. *Physics of Fluids* 10(1):223–236.

Burr, U., Barleon, L., Mack, K.-J., and Müller, U. (1999a). The effect of a horizontal magnetic field on liquid metal rayleigh-bénard convection. Technical Report FZKA 6277, Forschungszentrum Karlsruhe.

Burr, U., Barleon, L., Mack, K.-J., and Müller, U. (1999b). The effect of a vertical magnetic field on liquid metal rayleigh-bénard convection. Technical Report FZKA 6267, Forschungszentrum Karlsruhe.

Chandrasekhar, S. (1961). *Hydrodynamic and hydromagnetic stability*. New York: Dover Publications, Inc.

Chang, C., and Lundgren, S. (1961). Duct flow in magnetohydrodynamics. *Zeitschrift für angewandte Mathematik und Physik* XII:100–114.

Debray, F. (1997). Measurement of the onset of MHD-turbulence caused by a step in the electrical conductivity in the channel walls of GALINKA II and comparison with theoretical models. Technical Report FZKA 5972, Forschungszentrum Karlsruhe.

Frank, M., Barleon, L., and Müller, U. (1997). Experimentelle Untersuchung zweidimensionaler MHD-Turbulenz. Technical Report FZKA 6021, Forschungszentrum Karlsruhe. Diplomarbeit.

Garandet, J., Albousière, T., and Moreau, R. (1992). Buoyancy driven convection in a rectangular enclosure with a transverse magnetic field. *Int. J. Heat Mass Transfer* 35(4):741–748.

Gold, R. R. (1962). Magnetohydrodynamic pipe flow. Part 1. *Journal of Fluid Mechanics* 13:505–512. compare Shercliff.

Hartmann, J. (1937). Hg-Dynamics I Theory of the laminar flow of an electrically conductive liquid in a homogeneous magnetic field. *Det Kgl. Danske Videnskabernes Selskab. Mathematisk-fysiske Meddelelser.* XV(6):1–27.

Huges, W. F., and Young, F. J. (1966). *The electromagnetics*. John Wiley & Sons Inc.

Hunt, J. C. R., and Leibovich, S. (1967). Magnetohydrodynamic flow in channels of variable cross-section with strong transverse magnetic fields. *Journal of Fluid Mechanics* 28(Part 2):241–260.

Hunt, J. C. R., and Stewartson, K. (1965). Magnetohydrodynamic flow in rectangular ducts. *Journal of Fluid Mechanics* 23:563–581.

Hunt, J. C. R., and Williams, W. E. (1968). Some electically driven flows in magnetohydrodynamics. Part 1. Theory. *Journal of Fluid Mechanics* 31(4):705–722.

Hunt, J. C. R. (1965). Magnetohydrodynamic flow in rectangular ducts. *Journal of Fluid Mechanics* 21:577–590.

Khine, Y., and Walker, J. (1998). Thermoelectric magnetohydrodynamic effects during bridgman semiconductor crystal growth with a uniform axial magnetic field. *Journal of Crystal Growth* 183:159–158.

Lenhart, L., and McCarthy, K. (1991). Comparison of the core flow solution and the full solution for MHD flow. In *Proceedings of the Sixth Beer-Sheva International Seminar on MHD Flows and Turbulence, Jerusalem, 1990*. American Institute of Aeronautics and Astronautics, Inc. 482–499. Beer-Sheva.

Ludford, G. S. S. (1960). The effect of a very strong magnetic cross-field on steady motion through a slightly conducting fluid. *Journal of Fluid Mechanics* 10:141–155.

Lykoudis, P. S. (1996). Natural convection in a cubic enclosure in the presence of a horizontal magnetic field. *Journal of Heat Transfer* 118:215–218.

Ma, N., and Walker, J. S. (1995). Liquid-metal buoyant convection in a vertical cylinder with a strong vertical magnetic field and with nonaxisymmetric temperature. *Physics of Fluids* 7(8):2061–2071.

Ma, N., and Walker, J. S. (1996). Buoyant convection during the growth of compound semiconductors by the liquid-encapsulated Czochralski process with an axial magnetic field and with non-axisymmetric temperature. *Journal of Fluids Engineering* 118(8):155–159.

Molokov, S., and Bühler, L. (1994). Liquid metal flow in a U-bend in a strong uniform magnetic field. *Journal of Fluid Mechanics* 267:325–352.

Molokov, S. (1993). Fully developed liquid-metal flow in multiple rectangular ducts in a strong uniform magnetic field. *European Journal of Mechanics, B/Fluids* 12(6):769–787.

Molokov, S. (1994). Liquid metal flows in manifolds and expansions of insulating rectangular ducts in the plane perpendicular to a strong magnetic field. Technical Report KfK 5272, Kernforschungszentrum Karlsruhe.

Moon, T. J., Hua, T. Q., and Walker, J. S. (1991). Liquid-metal flow in a backward elbow in the plane of a strong magnetic field. *Journal of Fluid Mechanics* 227:273–292.

Moreau, R. (1990). *Magnetohydrodynamics*. Kluwer Academic Publisher.

Mößner, R. (1996). Dreidimensionale numerische Simulation von Naturkonvektionsströmungen unter dem Einfluß von Magnetfeldern. Technical Report FZKA 5748, Forschungszentrum Karlsruhe.

Murgatroyd, W. (1953). Experiments on magneto-hydrodynamic channel flow. *Phil. Mag.* 44:1348–1354.

Okada, K., and Ozoe, H. (1992). Experimental heat transfer rates of natural convcection of molten gallium suppressed under an external magnetic field in either the x, y, or z direction. *Journal of Heat Transfer* 114:107–114.

Reed, C. B., Picologlou, B. F., Hua, T. Q., and Walker, J. S. (1987). Alex results - A comparison of measurements from a round and a rectangular duct with 3-D code predictions. In *IEEE 12th Symposium on Fusion Engineering, Monterey, California, October 13-16*. 1267–1270.

Roberts, P. H. (1967). Singularities of Hartmann layers. *Proceedings of the Royal Society of London* 300(A):94–107.

Rosant, M. (1976). *Ecoulements hydromagnétiques turbulents en conduites rectangulaires*. Ph.D. Dissertation, Grenoble. see Moreau (1990), p149.

Series, R. W., and Hurle, D. T. J. (1991). The use of magnetic fields in semiconductor crystal growth. *Journal of Crystal Growth* 113:305–321.

Shercliff, J. A. (1953). *Proc.Camb.Phil.Soc.* 49:136.

Shercliff, J. A. (1962). Magnetohydrodynamic pipe flow Part 2. High Hartmann number. *Journal of Fluid Mechanics* 13:513–518. compare Gold 1962.

Sterl, A. (1990). Numerical simulation of liquid-metal MHD flows in rectangular ducts. *Journal of Fluid Mechanics* 216:161–191.

Stieglitz, R., and Molokov, S. (1997). Experimental study of magnetohydrodynamic flows in electrically coupled bends. *Journal of Fluid Mechanics* 343:1–28.

Stieglitz, R., Barleon, L., Bühler, L., and Molokov, S. (1996). Magnetohydrodynamic flow through a right-angle bend in a strong magnetic field. *Journal of Fluid Mechanics* 326:91–123.

Tabeling, P. (1982). Magnetohydrodynamic flows in rectilinear ducts of rectangular cross-section: the question of the corners. *Journal de Mécanique Théoretique et Appliquée* 1(1):25–38.

Temperley, D. J., and Todd, L. (1971). The effect of wall conductivity in magnetohydrodynamic duct flow at high Hartmann number. *Proc. Camb. Phil. Soc.* 69:337–351.

Ufland, Y. S. (1961). Hartman problem for a circular tube. *Soviet Physics Technical Physics* 5:1194–1196.

Walker, J. S. (1981). Magnetohydrodynamic flows in rectangular ducts with thin conducting walls. *Journal de Mécanique* 20(1):79–112.

Walker, J. (1998). Bridgman crystal growth with a strong, low-frequency, rotating magnetic field. *Journal of Crystal Growth* 192:318–327.

Chapter II

P.A. Davidson

MHD TURBULENCE AT LOW AND HIGH MAGNETIC REYNOLDS NUMBER

MHD Turbulence at Low and High Magnetic Reynolds Number

P A Davidson

Department of Engineering, University of Cambridge,
Trumpington Street, Cambridge, CB2 1PZ, U.K.

Abstract. This chapter discuses the influence of magnetic fields on turbulence, with particular emphasis on homogeneous turbulence away from boundaries. We start, however, by recalling certain features of conventional turbulence.

1 Conventional Turbulence

1.1 The Energy Cascade of Richardson and Kolmogorov

In any turbulent flow there exists a wide spectrum of eddy sizes. The way in which the kinetic energy of the flow is distributed across the different eddy sizes, and the manner in which these eddies interact, was first set out by Richardson and Kolmogorov. We describe some of their ideas here.

Suppose we have a statistically steady flow, say flow in a duct. Then the turbulent eddies are continually subject to viscous dissipation yet, by definition, the mean energy of the turbulence does not change. Thus the energy which is continually drained away by viscous forces must somehow be replenished by the mean flow. The usual way of quantifying this idea relies on dividing the flow, **u**, into two distinct parts, a mean component, $\overline{\mathbf{u}}$, and a turbulent component, $\mathbf{u}'$, and then examining the exchange of energy between the two. This exchange depends on the presence of so-called Reynolds stresses, τ_{ij}^R, and so we start by reminding you about these stresses.

When we time-average the Navier-Stokes equation in a turbulent flow the presence of the turbulence gives rise to additional stresses, $\tau_{ij}^R = -\rho\overline{u'_i u'_j}$, which act on the mean flow. (Here the overbar signifies a time average.) These Reynolds stresses give rise to a net force acting on the mean flow, $f_i = \partial\tau_{ij}^R / \partial x_j$, and if the rate of working of this force is negative, then the mean flow must lose energy to the turbulence. So when the turbulence is statistically steady, the viscous dissipation of the turbulent eddies is matched by the rate of production of turbulent energy by the Reynolds stresses. Symbolically we have,

$$P = \varepsilon \qquad (1a)$$

where ε is the dissipation per unit mass and P the rate of production of turbulent kinetic energy (also per unit mass).

Now the rate of working of f_i is $f_i\bar{u}_i$. When this is negative, energy is transferred from $\bar{\mathbf{u}}$ to $\mathbf{u}'$. Physically this corresponds to creation of turbulent eddies through some form of distortion of the mean flow. It turns out to be useful write $f_i\bar{u}_i$ in the form

$$f_i\bar{u}_i = \frac{\partial}{\partial x_j}\left[\bar{u}_i\tau_{ij}^R\right] - \tau_{ij}^R S_{ij} \qquad (2)$$

where S_{ij} is the strain-rate tensor,

$$S_{ij} = \frac{1}{2}\left[\frac{\partial \bar{u}_i}{\partial x_j} + \frac{\partial \bar{u}_j}{\partial x_i}\right]$$

The first term on the right of (2) is just the divergence of $\bar{u}_i\tau_{ij}^R$. In a finite, closed domain, or else in a statistically homogeneous turbulent flow, this term integrates to zero. More generally, it represents the work done by τ_{ij}^R on the boundary of some volume, V. Thus the *local* rate of transfer of mechanical energy to the turbulence is just $\rho P = \tau_{ij}^R S_{ij}$. This is called the *production of turbulent energy*. Clearly, in the absence of body forces, a finite strain-rate in the mean flow is necessary to keep the turbulence alive.

So in a steady-on-average flow then there is a continual energy transfer from the mean flow to the turbulence at a rate $\tau_{ij}^R S_{ij}$, and this energy is mostly held by the large, energetic eddies. Ultimately, of course, it is viscosity which destroys this energy. However, when Re is large, the viscous stresses acting on the large eddies are negligible, which raises the question of how the viscous forces can destroy the energy held by the large eddies. This leads to the idea of the energy cascade.

It is observed that any turbulent flow comprises structures – eddies - which have a wide range of sizes. That is to say, at any instant there is always a wide spectrum of length scales, velocity gradients, and so on. Richardson (and later Kolmogorov) suggested that the largest eddies, which are created by instabilities in the mean flow, are themselves subject to inertial instabilities and rapidly break up into yet smaller vortices. These smaller eddies then, in turn, become unstable and break up into even smaller vortices and so the process continues. In short, there is a *cascade* of energy from the largest eddies down to the smallest structures. This cascade is continually fed at the large scales by the production term $\rho P = \tau_{ij}^R S_{ij}$, while energy is removed at the small scales at a rate $\rho\varepsilon$.

Note, however, that viscosity plays almost no part in this cascade. That is, when Re is large the viscous forces acting on the large eddies are, by definition, negligible. The whole thing is simply driven by inertia. The cascade comes to a halt, however, when the eddies become so small that the Reynolds number based on the small-scale eddy size is of the order of unity. At this point the viscous forces become important and dissipation cuts in. So we may think of viscosity as providing a sink for energy at the very end of the cascade, but having no other effect.

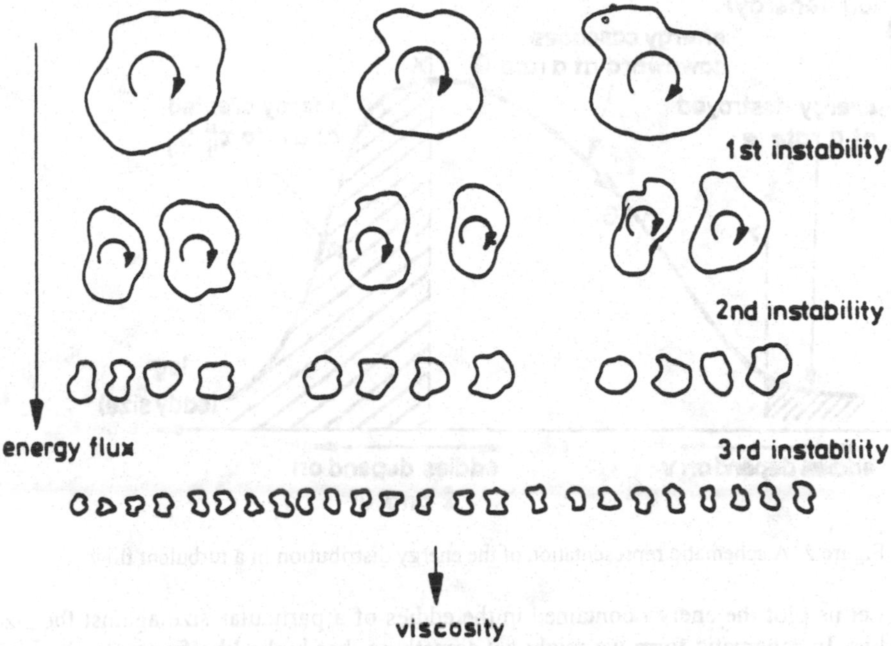

Figure 1 A cartoon of the energy cascade

It is useful to introduce some notation. Let l and u' be typical length and velocity scales for the larger eddies. For example, we might define u' via $(u')^2 = \overline{(u'_x)^2}$. Also, let v and η be typical velocity and length scales of the smallest structures in the flow. (The quantities v and η are known as the *Kolmogorov microscales*, whereas l is known as the *integral scale*.) Now we know that $v\eta/\nu \sim 1$. That is, the size of the small eddies is always such as to make the viscous forces an order-one quantity. Also, the energy dissipation rate per unit mass, which in a laminar flow is $-\nu(\nabla^2\mathbf{u})\cdot\mathbf{u}$, must be of order $\varepsilon \sim \nu v^2/\eta^2$. However, we have also seen that, in statistically steady turbulence, $\rho\varepsilon$ is equal to the rate at which energy is fed to the turbulence from the mean flow, $P = \tau_{ij}^R S_{ij} / \rho$. In fact, we can go further than this. If we are to avoid a build-up of eddies of a particular size, ε must equal the rate at which energy is past down the cascade at any point within that cascade. Let G be the rate at which energy is passed down the cascade. Then the generalisation of (1a) is,

$$P = G = \varepsilon \tag{1b}$$

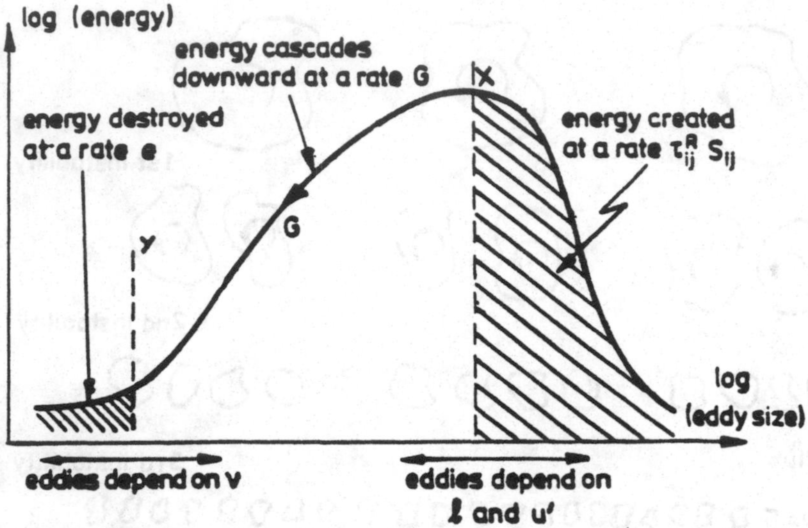

Figure 2 A schematic representation of the energy distribution in a turbulent flow

Let us plot the energy contained in the eddies of a particular size against the size of those eddies. In schematic form we might get something that looks like figure 2. Since there is dissipation only at the smallest scales G has to be the same at all points between X and Y, i.e. $G_X = G_Y$. Moreover, it is observed that the rate of extraction of energy from the large eddies is of the order of

$$G_X \sim (u')^3 / l \tag{3}$$

Physically, this implies that the large eddies break up on a time scale of their turn-over time.

We are now in a position to determine the size of the smallest eddies. In summery we know the following:

- It is observed that energy is extracted from the large scales and transferred to the energy cascade at a rate,

$$P = G_X \sim (u')^3 / l \tag{4}$$

- The cascade is inviscid except at the smallest scales and so, in steady-on-average turbulence

$$\varepsilon = G = P \tag{5,6}$$

- The smallest scales must satisfy

$$\varepsilon \sim \nu \mathrm{v}^2 / \eta^2, \qquad \mathrm{v}\eta / \nu \sim 1 \tag{7,8}$$

We may use (4)-(8) to express v and η in terms of l and u. After a little algebra we find,

$$\eta / l \sim (u'l / \nu)^{-3/4}, \qquad \mathrm{v}/ u' \sim (u'l / \nu)^{-1/4} \tag{9,10}$$

Since Re is invariably large these imply that v and η will be relatively small. For example, suppose that Re ~ 10^6 and l ~ 10cm. Then η ~ 0.003mm, which is very small indeed. There is, therefore, a large spectrum of eddy sizes in any turbulent flow. Moreover, equations (9) and (10) tell us that, as Re rises, so the differences between the largest and the smallest eddies become more marked. This is shown schematically in Figure 3 which compares two jets at different values of Re.

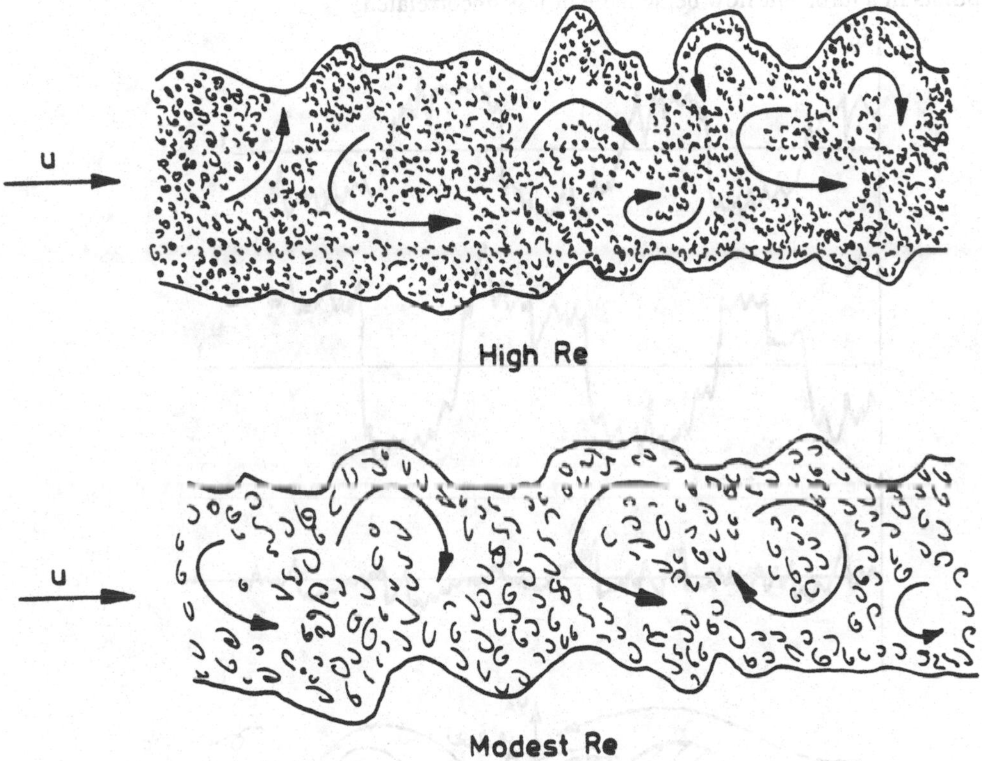

Figure 3. The effect of Re on the size of the smallest eddies.

1.2 The Karman-Howarth Equation

The arguments so far have been more than a little heuristic, and so we now turn to a more formal statistical description of turbulence. The common currency of turbulence theory is the *velocity correlation tensor*, sometimes called the velocity correlation function. The *second order velocity correlation tensor* is defined as

$$Q_{ij}(\mathbf{x},\mathbf{r}) = \overline{u'_i(\mathbf{x})u'_j(\mathbf{x}+\mathbf{r})} \tag{11}$$

It is not immediately obvious what Q_{ij} represents and so, to focus thoughts, let us take a specific example. Suppose we consider the turbulent wake behind a cylinder as shown in figure 4. Near the cylinder there are coherent vortices embedded in the turbulence, rather like a Karman vortex street. Consider two points, A and B, separated by $\mathbf{r} = r\,\mathbf{e}_y$. The correlation function Q_{yy} represents the degree to which the vertical velocities at A and B are correlated when averaged over time. Since the velocity fluctuations at A and B are statistically dependent (what is happening at A is roughly out of phase with what is happening at B) $Q_{yy}(r)$ is non-zero and negative. On the other hand, quite different things are happening at A and C in Figure 4, and so $Q_{yy} \approx 0$. We expect that $Q_{yy} \to \overline{u'^2_y}$ as $\mathbf{r} \to 0$ and $Q_{yy} \to 0$ as $\mathbf{r} \to \infty$, remote points in a turbulent flow being more or less uncorrelated.

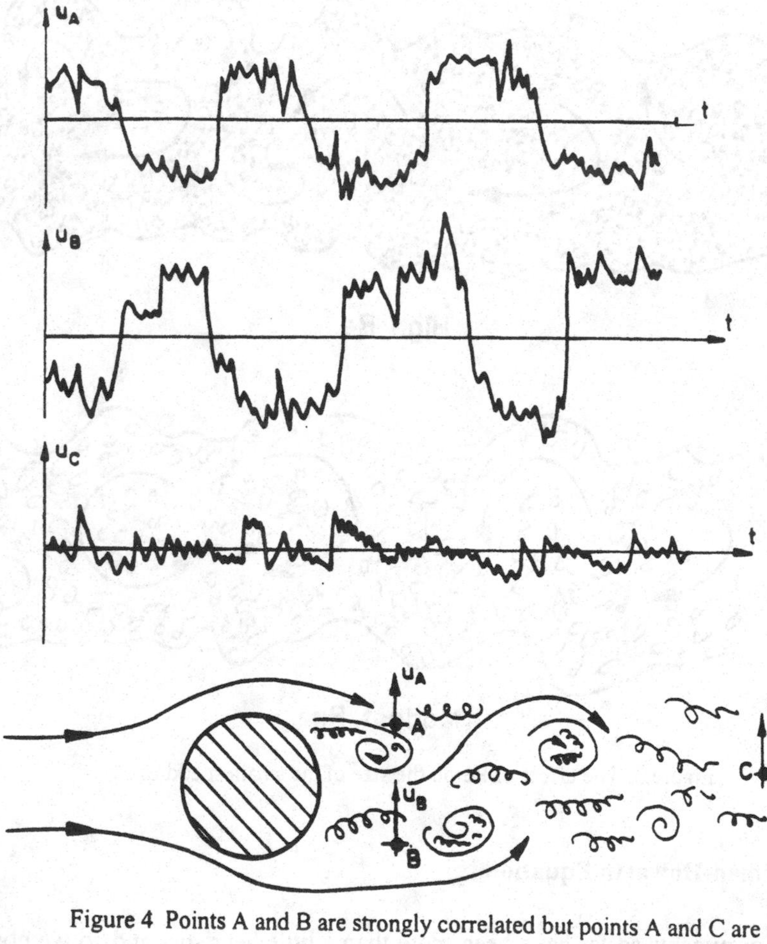

Figure 4 Points A and B are strongly correlated but points A and C are not.

We now restrict ourselves to an idealised form of turbulence. We consider turbulence which is statistically homogeneous and isotropic, and in which the mean velocity is zero. Since, in the absence of a mean shear, there is no mechanism for injecting energy into the turbulence, such a flow will always decay with time. Perhaps the easiest way to generate approximately homogeneous turbulence is to pass air uniformly through a mesh in a wind tunnel and adopt a frame of reference moving with the mean flow, as shown in figure 5.

Since the properties of the turbulence are now time-dependent we need to introduce a different means of taking averages. We rely on ensemble averages, i.e. an average over many realisations of the flow. This is represented by $\langle .. \rangle$. In homogeneous turbulence such an average is equivalent to a spatial average, while in statistically steady turbulence ensemble averages are equivalent to time averages.

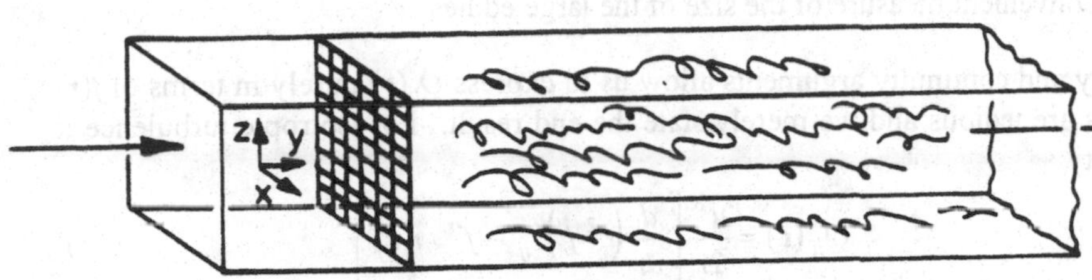

Figure 5. Grid turbulence in a wind tunnel

Note that, when the turbulence is homogeneous, Q_{ij} does not depend on **x**. Also, since the mean velocity (in a suitable frame of reference) is now taken to be zero, there is no need to use a prime to indicate a fluctuating velocity component. So the correlation function is now written as

$$Q_{ij}(\mathbf{r},t) = \langle u_i(\mathbf{x})u_j(\mathbf{x}+\mathbf{r})\rangle$$

This has the geometric property

$$Q_{ij}(\mathbf{r}) = Q_{ji}(-\mathbf{r}) \tag{12}$$

and is related to the kinetic energy per unit mass and Reynolds stress by

$$\tfrac{1}{2}\langle \mathbf{u}^2\rangle = \tfrac{1}{2}Q_{ii}(\mathbf{0}) \tag{13}$$

$$\tau_{ij}^R = -\rho Q_{ij}(\mathbf{0}). \tag{14}$$

We now introduce some additional notation. Let u be a characteristic turbulence velocity, defined by

$$u^2 = \langle u_x^2\rangle = \langle u_y^2\rangle = \langle u_z^2\rangle \tag{15}$$

and write $Q_{xx}(r)$ in the form

$$Q_{xx}(r\hat{\mathbf{e}}_x) = u^2 f(r) \tag{16}$$

The function f is known as the longitudinal correlation function. It is dimensionless, satisfies $f(0) = 1$, and is found to be positive. The integral scale, l, of the turbulence is often defined as

$$l = \int_0^\infty f(r)dr \tag{17}$$

which provides a convenient measure of the size of the large eddies.

Symmetry and continuity arguments allow us to express $Q_{ij}(\mathbf{r})$ purely in terms of $f(r)$ and $\mathbf{r}$. The details are tedious and we merely state the end result. For isotropic turbulence it may be shown that

$$Q_{ij}(\mathbf{r}) = \frac{u^2}{2r}\left[\frac{d}{dr}\left(r^2 f\right)\delta_{ij} - f' \ r_i r_j\right] \tag{18}$$

Next, we introduce the *third order velocity correlation function*,

$$S_{ijl}(\mathbf{r}) = \left\langle u_i(\mathbf{x})u_j(\mathbf{x})u_l(\mathbf{x}+\mathbf{r})\right\rangle$$

It too can be written in terms of a single scalar function, $k(r)$, defined as,

$$u^3 k(r) = \left\langle u_x^2(\mathbf{x})u_x(\mathbf{x}+r\hat{\mathbf{e}}_x)\right\rangle \tag{19}$$

The function k is known as the longitudinal triple velocity correlation function. Again, symmetry and continuity arguments may be used to show that, in homogeneous, isotropic, turbulence

$$S_{ijl} = u^3\left[\left(\frac{k - rk'}{2r^3}\right)r_i r_j r_l + \left(\frac{2k + rk'}{4r}\right)(r_i\delta_{jl} + r_j\delta_{il}) - \frac{k}{2r}r_l\delta_{ij}\right] \tag{20}$$

Let us now introduce some dynamics. Let $\mathbf{x}' = \mathbf{x} + \mathbf{r}$ and $\mathbf{u}' = \mathbf{u}(\mathbf{x}')$. (From now on, a prime will indicate a quantity at position $\mathbf{x}'$, rather than a fluctuating parameter.) Then we have

$$\frac{\partial u_i}{\partial t} = -\frac{\partial}{\partial x_k}(u_i u_k) - \frac{\partial}{\partial x_i}(p/\rho) + \nu\nabla_x^2 u_i$$

$$\frac{\partial u'_j}{\partial t} = -\frac{\partial}{\partial x'_k}(u'_j u'_k) - \frac{\partial}{\partial x'_j}(p'/\rho) + \nu\nabla_{x'}^2 u'_j$$

On multiplying the first of these by u'_j, and the second by u_i, adding the two and averaging, we find

$$\frac{\partial}{\partial t}Q_{ij} = -\left\langle u_i \frac{\partial u'_j u'_k}{\partial x'_k} + u'_j \frac{\partial u_i \partial u_k}{\partial x_k} \right\rangle - \frac{1}{\rho}\left\langle u_i \frac{\partial p'}{\partial x'_j} + u'_j \frac{\partial p}{\partial x_i} \right\rangle$$
$$+ \nu \left\langle u_i \nabla^2_{x'} u'_j + u'_j \nabla^2_x u_i \right\rangle \tag{21}$$

We now note that u_i is independent of $\mathbf{x}'$ while u'_j is independent of $\mathbf{x}$, and that $\partial / \partial x_i$ and $\partial / \partial x'_j$ operating on averages may be replaced by $-\partial / \partial r_i$ and $\partial / \partial r_j$ respectively. Expression (21) then simplifies to:

$$\frac{\partial Q_{ij}}{\partial t} = \frac{\partial}{\partial r_k}\left[S_{ikj} + S_{jki}\right] + 2\nu \nabla^2 Q_{ij} \tag{22}$$

Note that we have dropped the terms involving pressure since it may be shown that, in isotropic turbulence,

$$\left\langle p u'_j \right\rangle = 0 \, . \tag{23}$$

Finally, substituting for Q_{ij} and S_{ijk} in terms of the scalar functions $f(r)$ and $k(r)$ yields,

$$\boxed{\frac{\partial}{\partial t}\left[u^2 r^4 f(r)\right] = u^3 \frac{\partial}{\partial r}\left[r^4 k(r)\right] + 2\nu u^2 \frac{\partial}{\partial r}\left[r^4 f'(r)\right]} \tag{24}$$

We have arrived at the famous Karman-Howarth equation, which represents one of the central results in the theory of isotropic turbulence. Note that equation (24) may be integrated to give

$$\boxed{\frac{\partial}{\partial t}\int u^2 r^4 f(r) dr = \left[u^3 r^4 k(r)\right]_\infty + 2\nu \left[u^2 r^4 f'(r)\right]_\infty} \tag{25}$$

1.3 Loitsyansky's Integral and Kolmogorov's Law

Our intuition suggests that well-separated points in a turbulent flow should be statistically uncorrelated and so we expect $f(r)$ and $k(r)$ to decrease rapidly with distance. In fact, up until 1956 it was assumed that f and k decay transcendentally fast at large r. If this is indeed the case then (25) yields

$$\boxed{\frac{\partial}{\partial t}\int u^2 r^4 f(r)dr = 0} \tag{26a}$$

or

$$I = u^2 \int_0^\infty r^4 f(r)dr = \text{constant} \tag{26b}$$

I is known as Loitsyansky's integral. Kolmogorov took advantage of (26) to predict the rate of decay of energy in freely evolving, isotropic turbulence. That is, from (26) we have

$$I \sim u^2 l^5 = \text{constant} \tag{27}$$

Also, the large eddies tend to break up on a timescale of their turn-over time, and so

$$\frac{du^2}{dt} \sim -\frac{u^3}{l} \tag{28}$$

Combining (27) and (28) yields,

$$u^2 = u_0^2\left[1+(7/10)(u_0 t/l_0)\right]^{-10/7} \tag{29}$$

$$l \sim l_0\left[1+(7/10)(u_0 t/l_0)\right]^{2/7} \tag{30}$$

These are known as *Kolmogorov's decay laws*. (Here u_0 and l_0 are initial values of u and l.) It turns out that these predictions are reasonably in line with the experimental data, which typically give $l \sim t^{0.35}$ and $u^2 \sim t^{-1.26} \to t^{-1.34}$. All-in-all, it would seem that the experiments tend to support (26). There are two problems, however. First, if we are to believe (26), then we need some evidence that f and k decay exponentially, rather than algebraically, at large r. Second, we would really like some simple physical explanation for the invariance of I.

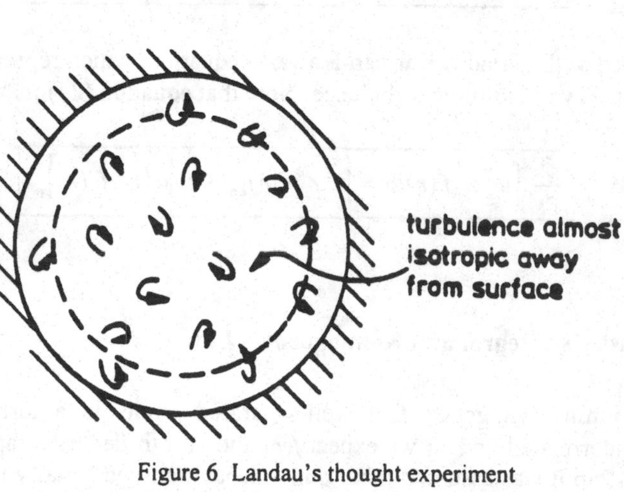

Figure 6 Landau's thought experiment

Landau resolved the second of these issues. He showed that, *provided f and k decay exponentially at large r*, then the invariance of I is a direct consequence of the conservation of angular momentum. Landau's argument centres around the thought experiment shown in Figure 6. Suppose the turbulent flow evolves in a large, closed sphere, whose radius R is much greater than the integral scale l. The global angular momentum of the turbulence is

$$\mathbf{H} = \int_V \mathbf{x} \times \mathbf{u} dV \tag{31}$$

and it turns out that the square of $\mathbf{H}$ can be written as

$$\mathbf{H}^2 = -\int_V \int_{V'} (\mathbf{x}' - \mathbf{x})^2 \mathbf{u} \cdot \mathbf{u}' dV dV'$$

We now ensemble average over each pair of points separated by a fixed distance $\mathbf{r} = \mathbf{x}' - \mathbf{x}$

$$\langle \mathbf{H}^2 \rangle = -\int \int \mathbf{r}^2 \langle \mathbf{u} \cdot \mathbf{u}' \rangle d^3\mathbf{r} dV$$

Next, we assume that $\langle \mathbf{u} \cdot \mathbf{u}' \rangle$ decays rapidly with $\mathbf{r}$ so that far-field contributions to the inner integral are small. In such a situation only those velocity correlations taken close to the surface are aware of the presence of the boundary and so to leading order in l/R we have

$$\boxed{\langle \mathbf{H}^2 \rangle / V = -\int \mathbf{r}^2 \langle \mathbf{u} \cdot \mathbf{u}' \rangle d^3\mathbf{r}} \tag{32}$$

Finally we note that (18) allows us to evaluate the integral on the left, which turns out to be $8\pi I$. So we have

$$\boxed{\langle \mathbf{H}^2 \rangle / V = 8\pi I} \tag{33}$$

It appears, therefor, that the invariance of I follows directly from conservation of angular momentum. According to Landau, then, Kolmogorov's decay law is a simple consequence of the conservation law (32). Note, however, that Landau had to make the same assumption as Loitsyansky and Kolmogorov: that remote points in the turbulence are statistically independent. It was this assumption which came to be questioned in later years.

The tide turned against Loitsiansky in 1956 when Batchelor showed that, in anisotropic turbulence, $k \sim r^{-4}$ as $r \to \infty$. If this is also true of isotropic turbulence then (25) gives

$$\frac{dI}{dt} = \left[u^3 r^4 k\right]_\infty \neq 0$$

In such a situation both Loitsyansky's and Landau's arguments fail. The reason for the relatively slow decline of k in anisotropic turbulence is rather subtle. It arises from the action of the pressure forces. A fluctuation in $\mathbf{u}$ at any one point in a flow gives rise to an associated pressure field, and this produces pressure forces, and hence accelerations, which fall off algebraically with distance from the source. Thus, because of pressure, a fluctuation in **u** at any one point is felt everywhere within the flow. Long-range pressure-velocity correlations result, and these appears as source terms in (21), causing the formation of long-range velocity correlations of the form $\langle \mathbf{u}\cdot\mathbf{u}'\rangle \sim r^{-5}$.

For *isotropic* turbulence, however, the symmetry is sufficiently strong to cause the direct effect of the long-range pressure forces to exactly cancel, as indicated by (23). Never-the-less, the pressure forces can still influence the triple correlations in isotropic turbulence and these, in turn, can influence the double correlations. In particular, it may be shown that, if $s = u_x^2 - u_y^2$,

$$\frac{d^2 I}{dt^2} = \frac{d}{dt}\left[u^3 r^4 k\right]_\infty = 3\int r^2 \langle ss'\rangle dr = J$$

(Davidson, 2001). In general, then, we would expect I to be time dependant. This is a direct result of the pressure forces which induce a $k_\infty \sim r^{-4}$ algebraic tail in the triple correlations and thus an r^{-6} tail in f_∞. Note, however, that we have yet to determine the magnitude of J. Interestingly, experiments suggests that J is extremely small. There are no direct measurements of J, but there is some indirect ways of assessing its size. First, there have been measurements of $u^2(t)$. Second, there exist measurements of Q_{ij} in the so-called final period of decay: do these show exponential or algebraic behaviour at large r?

It seems that the experiments tend to support Landau and Loitsyansky to they extent that they suggest that $J \approx 0$ in fully developed, isotropic turbulence. For example, the form of Q_{ij} at large r is exponential rather than algebraic in the final period of decay. Moreover, the measured decay rate of isotropic turbulence is not too far out of line with Kolmogorov's law. In 1960 Corrsin found $u^2 \sim t^{-n}$ where n lies in the range 1.2→ 1.4 with an average value of 1.26. Later Lumley found $u^2 \sim t^{-1.34}$ and $l \sim t^{0.35}$.

In summary, then, it would seem likely that the Landau-Loitsyansky equation

$$\langle \mathbf{H}^2\rangle / V = 8\pi I = const.$$

is approximately valid in fully developed, isotropic turbulence.

2. MHD Turbulence

We now turn to MHD turbulence. How should we attack the problem? The obvious thing to do is to manipulate the equations of motion into a generalised version of (21), taking into account the Lorentz force **J×B**. This is tedious but routine and the end result is a generalised Karman-Howarth equation of the form

$$\frac{\partial Q_{ij}}{\partial t} = \frac{\partial}{\partial r_k}\left[S_{ikj} - \langle b_i b_k u_j'\rangle/(\rho\mu) + S_{jki} - \langle b_j b_k u_i'\rangle/(\rho\mu) \right] +$$

$$\frac{1}{\rho}\left[\langle (\mathbf{J}\times\mathbf{B}_0)_i u_j' + (\mathbf{J}'\times\mathbf{B}_0)_j u_i \rangle \right] + \frac{1}{\rho}\left[\frac{\partial}{\partial r_i}\langle pu_j'\rangle + \frac{\partial}{\partial r_j}\langle pu_i'\rangle\right] + 2\nu\nabla^2 Q_{ij} \qquad (34)$$

Here we have written the total magnetic field as $\mathbf{B} = \mathbf{B}_0 + \mathbf{b}$, where $\mathbf{B}_0$ is some imposed magnetic field (which might be zero) and $\mathbf{b}$ is the so-called induced field associated with currents locally generated by the turbulence interacting with **B**. But what does this equation tell us? We know that Ohmic heating enhances the decay of turbulence by converting mechanical energy into heat, and it is observed that an imposed magnetic field tends to elongate the turbulent eddies along the field lines, yet none of this is evident from (34). It turns out that an indirect attack on the problem is much more fruitful and we take our lead from Landau. We start by looking at the damping of a simple eddy by an imposed magnetic field, the idea being that turbulence is simply an ensemble of eddies. We then go on to repeat Landau's analysis, adapted to MHD turbulence.

2.1 The damping of a simple eddy.

Let us consider a simple example designed to bring out the tendency for vorticity and momentum to diffuse along magnetic field lines. For simplicity we shall take R_m to be small, $\mathbf{B}_0$ to be uniform, and the initial velocity field to be an axisymmetric, swirling vortex, $\mathbf{u} = (0, \Gamma/r, 0)$ in (r, θ, z) coordinates. (Γ is the angular momentum per unit mass.) Since R_m is small the induced field **b** is negligible and Ohm's law simplifies to

$$\mathbf{J} = \sigma(-\nabla V + \mathbf{u}\times\mathbf{B}_0)\ , \quad \nabla\cdot\mathbf{J} = 0$$

We shall take $\mathbf{B}_0$ to be parallel to the z-axis and drop the subscript on $\mathbf{B}_0$ since $\mathbf{b} \ll \mathbf{B}$.

At $t = 0$ $\Gamma(r,z)$, is assumed to be confined to a spherical region of size δ. Now the axial gradients in Γ will tend to induce a poloidal component of motion, $\mathbf{u}_p = (u_r, 0, u_z)$. That is, if Γ is a function of z then the centrifugal force, $(\Gamma^2/r^3)\hat{\mathbf{e}}_r$, is rotational and cannot be balanced by a pressure gradient. A poloidal component of motion then results which complicates the problem. However, in the interests of simplicity, we shall take $\mathbf{J}\times\mathbf{B} \gg \mathbf{u}\cdot\nabla\mathbf{u}$, which is equivalent to specifying that the magnetic damping time,

$$\tau = \left(\sigma B^2/\rho\right)^{-1}$$

is much less than the inertial time-scale δ/u. We may then neglect $\mathbf{u}_p$ for times of order τ, which is long enough to see what **B** is doing to the vortex.

Let us now determine the induced current, **J**, and hence the Lorentz force which acts on the vortex. The term $\mathbf{u}\times\mathbf{B}$ in Ohm's law gives rise to a radial component of current, J_r. However, the current lines must form closed paths and so an electrostatic potential, $V(r,z)$, is established, whose primary function is to ensure that the **J**-lines close. The distribution of V is fixed by the divergence of Ohm's law $\nabla^2 V = \mathbf{B}\cdot\nabla\times\mathbf{u}$. It drives an axial component of current, thus allowing **J** to form closed current paths in the r-z. The current density, on the other hand, is fixed by the curl of Ohm's law,

$$\nabla\times\mathbf{J} = \sigma\mathbf{B}\cdot\nabla\mathbf{u} \tag{35}$$

and since **J** is solenoidal we can introduce a vector potential defined as follows

$$\mathbf{J} = \nabla \times \left[(\phi / r)\hat{\mathbf{e}}_\theta\right] = \left(-\frac{1}{r}\frac{\partial \phi}{\partial z}, 0, \frac{1}{r}\frac{\partial \phi}{\partial r}\right)$$

Combining this expression with the curl of Ohm's law yields,

$$\nabla_*^2 \phi = -\sigma B \frac{\partial \Gamma}{\partial z}, \qquad \nabla_*^2 \equiv \frac{\partial^2}{\partial z^2} + r\frac{\partial}{\partial r}\left(\frac{1}{r}\frac{\partial}{\partial r}\right).$$

The Lorentz force per unit mass, $\mathbf{F} = -(J_r B / \rho)\hat{\mathbf{e}}_\theta$, can now be written in terms of Γ:

$$rF_\theta = \frac{B}{\rho}\frac{\partial \phi}{\partial z} = -\frac{1}{\tau}\frac{\partial^2}{\partial z^2}\left(\nabla_*^{-2}\Gamma\right).$$

Next, we need the azimuthal equation of motion . Ignoring the poloidal motion and viscous stresses this is simply

$$\frac{D\Gamma}{Dt} = \frac{\partial \Gamma}{\partial t} = rF_\theta = \frac{B}{\rho}\frac{\partial \phi}{\partial z} = -\frac{1}{\tau}\frac{\partial^2}{\partial z^2}\left(\nabla_*^{-2}\Gamma\right)$$

and we see immediately that the global angular momentum, H, of the vortex is conserved,

$$\frac{dH}{dt} = \frac{d}{dt}\int \Gamma dV = \frac{B}{\rho}\int \nabla \cdot \left[\phi \hat{\mathbf{e}}_z\right] dV = 0.$$

However, kinetic energy is continually converted into heat, via Ohmic heating, in accordance with ,

$$\frac{dE}{dt} = -\frac{1}{\rho\sigma}\int J^2 dV \quad , \quad E = \tfrac{1}{2}\int u^2 dV$$

So the vortex must somehow arrange for H to be conserved as E falls. Let us see how this is achieved. Let l_r and l_z be characteristic radial and axial length scales for the vortex. At $t = 0$ we have $l_r = l_z = \delta$, and we shall suppose that l_r remains of order δ throughout the life of the vortex. Then (35) allows us to estimate the magnitude of **J**, from which,

$$\frac{dE}{dt} \sim -\frac{1}{\tau}\left(\frac{\delta}{l_z}\right)^2 E, \qquad E \sim \Gamma^2 l_z \tag{36}$$

However we also require,

$$H \sim \Gamma \delta^2 l_z = \text{constant} \tag{37}$$

Clearly l_z must increase with time since otherwise E would decay exponentially on a timescale of τ, which contradicts the conservation of H. The only way of satisfying both of the equation above is to have Γ and l_z scale as,

$$\Gamma \sim \Gamma_0 (t/\tau)^{-1/2} \quad , \quad l_z \sim \delta (t/\tau)^{1/2} \tag{38}$$

which, in turn, requires $E \sim (t/\tau)^{-1/2}$. So the vortex evolves from a sphere to an cigar-like shape on a timescale of τ.

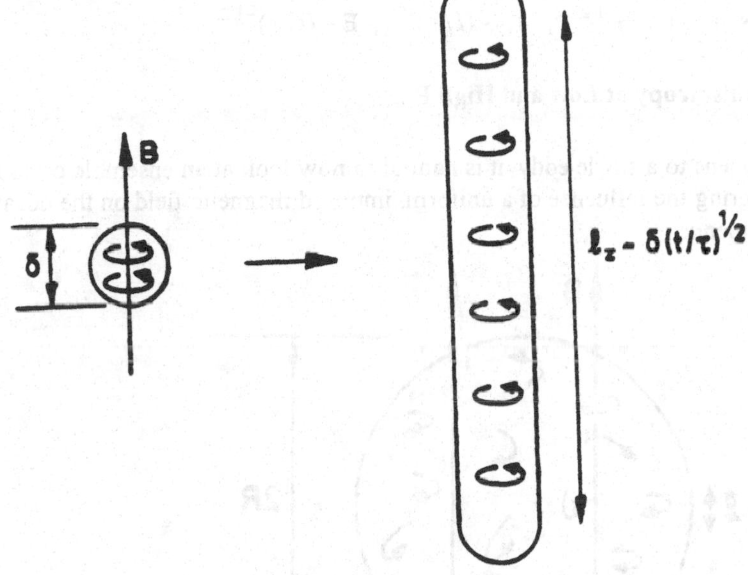

Figure 7. Turbulence evolving under the influence of an imposed magnetic field

It seems that angular momentum and vorticity diffuse along the magnetic field lines with a diffusivity of $\alpha_B \sim \delta^2/\tau$. (Recal that the diffusion rate in a typical problem is $l \sim \sqrt{\alpha t}$.) We might picture this diffusion process as a spiralling up of the magnetic field lines, which then slowly unwind, propagating angular momentum along the z-axis. Infact, it is possible to show that this pseudo-diffusion is the last vestige of Alfvèn wave propagation at low R_m.

The estimates above may be confirmed by exact analysis. The simplest method is to use the cosine-Hankel transform, defined by,

$$F(u_\theta) = U(k_r, k_z) = 4\pi \int_0^\infty \int_0^\infty [u_\theta] J_1(k_r r) \cos(k_z z) r dr dz$$

The transform of our equation of motion is then,

$$\frac{\partial U}{\partial t} = -\left[\cos^2 \alpha\right]\frac{U}{\tau} \quad , \cos\alpha = k_z / k$$

Solving for U and performing the inverse transform gives,

$$\Gamma = \frac{r}{2\pi^2}\int_0^\infty\int_0^\infty \left[U_0 \exp\left(-\cos^2 \alpha \ (t/\tau)\right)\right] J_1(k_r r)\cos(k_z z) k_r dk_r dk_z$$

where $U_0 = F(u_\theta)$ at $t = 0$. For $t >> \tau$ this takes the form $\Gamma = (t/\tau)^{-1/2} T(r, z/(t/\tau)^{1/2})$ which confirms the estimates $\Gamma \sim (t/\tau)^{-1/2}$, $l_z \sim (t/\tau)^{1/2}$, $E \sim (t/\tau)^{-1/2}$.

2.2 The Growth of Anisotropy at Low and High R_m

Having seen what happens to a single eddy it is natural to now look at an ensemble of eddies. Let us start by considering the influence of a uniform, imposed magnetic field on the decay of initially isotropic turbulence.

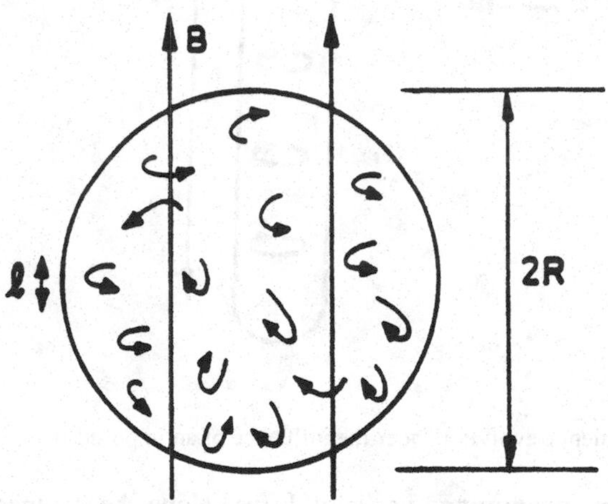

Figure 8. Turbulence evolving under the influence of an imposed magnetic field.

Suppose that a conducting fluid is held in a large, insulated sphere of radius R, which in turn sits in a uniform, imposed field $\mathbf{B}_0$. We imagine that, at $t = 0$, the fluid is mechanically stirred and then left to itself. No restriction is placed on the value of R_m, or on the size of the interaction parameter, $N = \sigma B_0^2 l / \rho u$. (Here l is the integral scale of the turbulence, assumed much less than R.) As before, the questions which interest us are:

(i) how fast does the energy decay?

(ii) can we characterize the anisotropy introduced into the turbulence by $\mathbf{B}_0$?

In view of the central role played by angular momentum in the evolution of a single vortex, it seems natural to once again focus attention on **H**. We start by noting that the torque exerted on the fluid by the Lorentz force is

$$\mathbf{T} = \int \mathbf{x} \times (\mathbf{J} \times \mathbf{B}_0) dV + \int \mathbf{x} \times (\mathbf{J} \times \mathbf{b}) dV$$

However, the second integral on the right is zero since a closed system of currents produce zero net torque when they interact with their self field, **b**. Next we transform the first integral using the relationship

$$2\mathbf{x} \times [\mathbf{J} \times \mathbf{B}_0] = [\mathbf{x} \times \mathbf{J}] \times \mathbf{B}_0 + \nabla \cdot [(\mathbf{x} \times (\mathbf{x} \times \mathbf{B}_0))\mathbf{J}] \qquad (39)$$

to give an expression for **T** in terms of the global dipole moment, **m**,

$$\mathbf{T} = \left\{ \frac{1}{2} \int (\mathbf{x} \times \mathbf{J}) dV \right\} \times \mathbf{B}_0 = \mathbf{m} \times \mathbf{B}_0$$

It follows that the global angular momentum evolves according to

$$\rho \frac{d\mathbf{H}}{dt} = \mathbf{T} + \mathbf{T}_\nu = \mathbf{m} \times \mathbf{B}_0 + \mathbf{T}_\nu \qquad (40a)$$

where $\mathbf{T}_\nu$ is the viscous torque exerted by the boundary. We are primarily interested in the influence of **B** on the turbulence and so let us, for the moment, take the fluid to be inviscid. (We shall put viscosity back in a moment.) Then,

$$\rho \frac{d\mathbf{H}}{dt} = \mathbf{T} = \mathbf{m} \times \mathbf{B}_0 \qquad (40b)$$

and it follows immediately that $\mathbf{H}_{//}$ is conserved. This, in turn, yields a lower bound on the total energy of the system via Schwarz inequality,

$$E = E_\mathbf{b} + E_\mathbf{u} > \mathbf{H}_{//}^2 \left(2 \int \mathbf{x}_\perp^2 dV \right)^{-1} \qquad (41)$$

Yet we know that the total energy declines due to Joule dissipation in accordance with,

$$\frac{dE}{dt} = -\frac{1}{\sigma\rho} \int_{V_R} \mathbf{J}^2 dV \qquad (42)$$

$$\rho E = \int_{V_R} \frac{\rho \mathbf{u}^2}{2} dV + \int_{V_\infty} \frac{b^2}{2\mu} dV$$

The only possibility of satisfying both (41) and (42) is for the turbulence to evolve to a state in which $\mathbf{J} = 0$, yet E_u is non zero (to satisfy (41)). However, if $\mathbf{J} = 0$ then Ohm's law requires $\mathbf{E} = -\mathbf{u} \times \mathbf{B}_0$, while Faraday's law tells us that $\nabla \times \mathbf{E} = \mathbf{0}$. It follows that, at large times, $(\mathbf{B}_0 \cdot \nabla)\mathbf{u} = \mathbf{0}$. The final state is therefore strictly two-dimensional, of the form

$$\mathbf{u}_\perp = \mathbf{u}_\perp(\mathbf{x}_\perp), \qquad \mathbf{u}_{//} = \mathbf{0}.$$

In short, the turbulence ultimately reaches a state which consists of one or more columnar eddies aligned with $\mathbf{B}_0$.

At low R_m this transition will occur on the timescale of $\tau = \left(\sigma B_0^2 / \rho\right)^{-1}$. We may show this as follows. At low R_m, Ohm's law simplifies to

$$\mathbf{J} = \sigma\left(-\nabla V + \mathbf{u} \times \mathbf{B}_0\right)$$

and so the dipole moment, $\mathbf{m}$, is given by

$$\mathbf{m} = \tfrac{1}{2}\int \mathbf{x} \times \mathbf{J} dV = (\sigma / 2)\int \mathbf{x} \times (\mathbf{u} \times \mathbf{B}_0) dV = (\sigma / 4)\mathbf{H} \times \mathbf{B}_0$$

Our angular momentum equation now becomes

$$\frac{d\mathbf{H}}{dt} = -\frac{\mathbf{H}_\perp}{4\tau} \tag{43}$$

Evidently $\mathbf{H}_\perp$ declines exponentially on a timescale of 4τ while, as expected, $\mathbf{H}_{//}$ is conserved.

Although we have found it convenient to frame the discussion in terms of a confined domain, it is clear that this elongation of the eddies should be independent of the existance of the remote boundaries. This is consistent with the discussion in the previous section where we saw that, at low R_m, isolated vortices in an infinite domain elongate in the direction of $\mathbf{B}_0$ at a rate $l_{//} / l_\perp \sim (t/\tau)^{1/2}$ while their energy declines as $u^2 \sim (t/\tau)^{-1/2}$.

Of course, in these arguments we have ignored ν and hence the process of energy removal via the energy cascade. In reality, for a finite ν, the predicted growth of anisotropy will occur only if the turbulence lives for long enough and this, in turn, requires $N \geq \sim 1$. However, the introduction of viscous forces does not influence the conservation of $\mathbf{H}_{//}$ since the viscous torque in (40a) has negligible influence as long as $R >> l$.

Now the conservation of $\mathbf{H}_{//}$ is important because, following Landau, it yields

$$\left\langle \mathbf{H}_{//}^2 \right\rangle = -\int \int \mathbf{r}_\perp^2 \left\langle \mathbf{u}_\perp \cdot \mathbf{u}'_\perp \right\rangle d^3\mathbf{r}\, d^3\mathbf{x} = \text{constant} \tag{44}$$

where $\mathbf{r} = \mathbf{x}' - \mathbf{x}$. If we can ignore Batchelor's long-range statistical correlations then we have the Loitsyansky-like invariant

$$\left\langle \mathbf{H}_{//}^2 \right\rangle / V = -\int \mathbf{r}_\perp^2 \langle \mathbf{u}_\perp \cdot \mathbf{u}'_\perp \rangle d^3\mathbf{r} = \text{constant} \tag{45}$$

Actually, this invariant can also be obtained by integrating the generalised Karman-Howarth equation

$$\frac{\partial Q_{ij}}{\partial t} = \frac{\partial}{\partial r_k}\left[S_{ikj} - \left\langle b_i b_k u'_j \right\rangle / (\rho\mu) + S_{jki} - \left\langle b_j b_k u'_i \right\rangle / (\rho\mu) \right] +$$

$$\frac{1}{\rho}\left[\left\langle (\mathbf{J}\times\mathbf{B}_0)_i u'_j + (\mathbf{J}'\times\mathbf{B}_0)_j u_i \right\rangle \right] + \frac{1}{\rho}\left[\frac{\partial}{\partial r_i}\left\langle p u'_j \right\rangle + \frac{\partial}{\partial r_j}\left\langle p u'_i \right\rangle \right] + 2\nu\nabla^2 Q_{ij}$$

and assuming all correlations are exponentially small at large r. However Landau's method is to be preferred since it exposes the underlying nature of the invariant. Note that (45) is valid for any N, provided that the long-range statistical correlations are weak. It is also valid for any R_m. We may use (45), in the spirit of Kolmogorov, to predict the rate of decay of energy.

2.3 Decay Laws at Low R_m

We now restrict ourselves to low values of R_m and seek the MHD equivalent of Kolmogorov's decay laws (29) and (30). These are based on the estimates

$$\frac{du^2}{dt} \sim -\frac{u^3}{l}, \qquad \int \mathbf{r}^2 \langle \mathbf{u}\cdot\mathbf{u}' \rangle d^3\mathbf{r} \sim u^2 l^5 = \text{const}$$

and so we require the MHD analogues of these expressions. In MHD turbulence the kinetic energy declines due to both Joule dissipation and viscous stresses,

$$\frac{dE}{dt} = -\frac{1}{\sigma}\int \mathbf{J}^2 dV - \rho\nu \int \omega^2 dV \tag{46}$$

Now let us suppose that the energy cascade proceeds as usual, on a timescale of l/u. Then the MHD analogue of (28) is

$$\frac{du^2}{dt} = -\frac{u^3}{l_\perp} - \left(\frac{l_\perp}{l_{//}}\right)^2 \frac{u^2}{\tau} \tag{47}$$

Here $l_\perp$ and $l_{//}$ represent suitably defined transverse and longitudinal integral scales. (The magnitude of the Joule dissipation in (47), $\langle \mathbf{J}^2 \rangle / \rho\sigma$, has been estimated using the curl of the low-R_m form of Ohm's law, (35), which yields $J \sim \sigma B_0 u\,(l_{//}/l_\perp)$.)We now need the analogue of (27). This is provided by our conservation law (45):

$$u^2 l_{//} l_\perp^4 = \text{constant} \tag{48}$$

However, we are not yet in a position to estimate the decline in energy. That is, because of the anisotropy of MHD turbulence, we have three, rather than two, unknowns: u, $l_{//}$, $l_\perp$. We need a third relationship if we are to predict the evolution of the flow. This comes from the fact that $l_{//}/l_\perp = 1$ if N is small and obeys (38) if N is large, i.e. $l_{//}/l_\perp \sim (t/\tau)^{1/2}$. Both cases are captured by the expression

$$\frac{d}{dt}\left(\frac{l_{//}}{l_\perp}\right)^2 = \frac{2}{\tau} \tag{49}$$

and we shall use (49) to interpolate between the two limits. Expressions (47) → (49) represent a closed system for u, $l_{//}$ and $l_\perp$. These are readily integrated to yield

$$\boxed{u^2/u_0^2 = \hat{t}^{-1/2}\left[1+(7/15)\left(\hat{t}^{3/4}-1\right)N_0^{-1}\right]^{-10/7}}$$

$$\boxed{l_\perp/l_0 = \left[1+(7/15)\left(\hat{t}^{3/4}-1\right)N_0^{-1}\right]^{2/7}}$$

$$\boxed{l_{//}/l_0 = \hat{t}^{1/2}\left[1+(7/15)\left(\hat{t}^{3/4}-1\right)N_0^{-1}\right]^{2/7}}$$

where N_0 is the initial value of $N = l_\perp/u\tau$ and $\hat{t} = 1+2(t/\tau)$. The high and low-N results given by (29), (30), and (38) are special cases of these. Note that, in general u^2, $l_{//}$ and $l_\perp$ do not obey simple power laws. However, for the special case of $N_0 = 7/15$ we have,

$$u^2 \sim (t/\tau)^{-11/7},\ l_{//}/l_0 \sim (t/\tau)^{5/7},\ l_\perp \sim (t/\tau)^{3/14}$$

Indeed the $u^2 \sim (t/\tau)^{-11/7}$ behaviour is a good approximation to solutions of (47) →(49) for all values of N around 1 and this compares favourably with laboratory experiments performed at N ~ 1, which show $u^2 \sim (t/\tau)^{-1.6}$.

In summary, then, eddies tend to elongate in the direction of B_0, causing $l_{//}$ to grow faster than $l_\perp$. There are several related explanations for the growth of $l_{//}$ given in the literature. The simplest explanation relates to Alfven waves. It is readily confirmed that the dispersion relationship for small-amplitude Alfven waves takes a curious form in the limit of low R_m. Instead of obtaining conventional wave motion we find a diffusive process in which energy and vorticity spreads along the **B**-lines at a rate $(t/\tau)^{1/2}$.

2.4 The Intensification of a Magnetic Field at High R_m

We now consider high-R_m turbulence. We start by considering forced, steady-on-average turbulence and seek to determine the conditions under which a small 'seed' field, present in the fluid at $t = 0$, will grow or decay. A simple argument, proposed by Batchelor, suggests that a seed field will grow if $\lambda < \nu$ and decay if $\lambda > \nu$. (Here λ is the magnetic diffusivity, $(\sigma\mu)^{-1}$.) The argument proceeds as follows. The ultimate fate of the seed field is determined by the balance between the Ohmic dissipation and the random stretching of the flux tubes by **u**, which tends to increase $<\mathbf{B}^2>$. Also, $\boldsymbol{\omega}$ and **B** are governed by similar equations:

$$\frac{\partial \omega}{\partial t} = \nabla \times (\mathbf{u} \times \omega) + \nu \nabla^2 \omega \tag{51}$$

$$\frac{\partial \mathbf{B}}{\partial t} = \nabla \times (\mathbf{u} \times \mathbf{B}) + \lambda \nabla^2 \mathbf{B} \tag{52}$$

Now suppose that $\lambda = \nu$. Then there exists a solution of the form $\mathbf{B} = \text{constant} \times \boldsymbol{\omega}$. Thus, if $<\omega^2>$ is steady, so is $<\mathbf{B}^2>$, and it follows that the flux tube stretching and Ohmic dissipation just balance. If $\lambda > \nu$, however, we would expect enhanced Ohmic dissipation and a fall in $<\mathbf{B}^2>$, while $\lambda < \nu$ should lead to the spontaneous growth of the seed field.

Unfortunately, these simple arguments are imperfect. The problem is that the analogy between **B** and $\boldsymbol{\omega}$ is not exact. If the turbulence is to be statistically steady then a forcing term must appear in the vorticity equation representing mechanical stirring, yet the corresponding term is absent in the induction equation. So the exact conditions under which $<\mathbf{B}^2>$ will spontaneously grow in forced, high-R_m turbulence are still unknown.

We now turn to freely decaying turbulence. Here there are arguments to suggest that there is an *inverse cascade* of the magnetic field. That is to say, the integral scale for **B** grows as the flow evolves. The arguments rest on the approximate conservation of magnetic helicity, **A·B**, which, in turn, relies on the equation,

$$\frac{D}{Dt}(\mathbf{A} \cdot \mathbf{B}) = \nabla \cdot [(\mathbf{u} \cdot \mathbf{A})\mathbf{B}] - \sigma^{-1}[2\mathbf{J} \cdot \mathbf{B} + \nabla \cdot (\mathbf{J} \times \mathbf{A})] \ , \quad \mathbf{B} = \nabla \times \mathbf{A} \tag{53}$$

This tells us that magnetic helicity is globally conserved when $\lambda = 0$ and should be almost conserved if λ is small. We now assume that the turbulence is statistically homogeneous so that the divergences of averaged quantities disappear:

$$\frac{d}{dt}[\langle \mathbf{A} \cdot \mathbf{B} \rangle] = -2\langle \mathbf{J} \cdot \mathbf{B} \rangle / \sigma \tag{54}$$

In addition, we note that there is decline of energy through viscous and Ohmic dissipation,

$$\frac{d}{dt}\left[\frac{1}{2}\rho\langle u^2 \rangle + \langle B^2 \rangle / 2\mu\right] = -\rho\nu\langle \omega^2 \rangle - \langle J^2 \rangle / \sigma \tag{55}$$

Let us rewrite (54) and (55) as,

$$\frac{dH_B}{dt} = -2\langle \mathbf{J} \cdot \mathbf{B} \rangle / \sigma \tag{56}$$

$$\frac{dE}{dt} = -\rho\nu\langle\omega^2\rangle - \langle J^2\rangle/\sigma \tag{57}$$

Next we note that the Schwarz inequality tells us

$$\left\{\int \mathbf{J}\cdot\mathbf{B}dV\right\}^2 \le \int J^2 dV \int B^2 dV \tag{58}$$

which yields

$$|\langle\mathbf{J}\cdot\mathbf{B}\rangle|/\sigma \le (2\mu/\sigma)^{1/2}\left[\dot{E}|E\right]^{1/2}$$

This allows us to place an upper bound on the rate of destruction of magnetic helicity:

$$|\dot{H}_B|/\mu \le (8\lambda)^{1/2}|\dot{E}|^{1/2} E^{1/2} \tag{59}$$

We are interested in high-Rm turbulence and so the next step is to consider what happens when $\sigma \to \infty$, ($\lambda \to 0$). In the process, however, we assume that $\dot{E}$ remains finite. It follows that, in the limit $\lambda \to 0$, H_B is conserved. Thus, for small λ, we have the destruction of energy subject to the conservation of magnetic helicity. This presents us with a simple variational problem. Minimising E subject to the conservation of H_B in a bounded domain yields $\mathbf{u} = 0$, $\nabla\times\mathbf{B} = \alpha\mathbf{B}$. Here the eigenvalue α is of the order of $\alpha^{-1} \sim l_{\text{domain}}$. Evidently, **B** ends up with a length scale comparable with the domain size.

Suggested Reading

Conventional Turbulence

L D Landau & E M Lifshitz, *Course of theoretical physics, vol. 6, Fluid mechanics.* 1st Edition, 1959. Butterworth-Heinemann Ltd.

H Tennekes & J L Lumley, *A first course in turbulence,* 1972, The MIT Press.

J O Hinze, *Turbulence,* 1959. McGraw-Hill Co.

G K Batchelor, *The theory of homogeneous turbulence,* 1953.

M Lesieur, *Turbulence in Fluids,* 1990. Kluwer Acad. Pub.

MHD Turbulence

D Biskamp, *Nonlinear magnetohydrodynamics,* 1993. Cambridge University Press.

P A Davidson , *An Introduction to Magnetohydrodynamics,* 2001. Cambridge University Press.

Chapter III

M.R.E. Proctor

MHD AT LARGE MAGNETIC REYNOLDS NUMBER

MHD at large Magnetic Reynolds number

M.R.E. PROCTOR

Department of Applied Mathematics and Theoretical Physics
University of Cambridge, Silver Street, Cambridge CB3 9EW, U.K.

Abstract. This chapter gives an overview of the effects of large magnetic Reynolds number on the MHD of convecting fluids. Topics discussed include flux expulsion, MHD waves, and magnetoconvection.

1 Introduction

While most of the chapters in this volume deal with laboratory and industrial MHD, in which inductively induced fields are always small compared with the imposed field, this chapter deals with problems of astrophysical interest, in which the scale of the domain is so large, or the conductivity so low, that induction dominates. The dimensionless number describing the relative effects of induction and diffusion is of course the *magnetic Reynolds number* $R_m \equiv \mathcal{U}\mathcal{L}/\eta$, where $\mathcal{U}$, $\mathcal{L}$ are appropriate velocity and length scales for the flow. With conductivities of the order of $1\mathrm{m}^2\mathrm{s}^{-1}$ for liquid metals, it is only recently that laboratory devices have achieved values of R_m significantly greater than unity. In the Earth's core, however, we have self-sustained fields so that according to anti-dynamo theorems we expect R_m to be significantly greater than unity, in the Sun length scales are so vast that R_m is huge.

What new processes are produced by having $R_m \gg 1$? There are both kinematic and dynamical effects. On the kinematic side, we have the phenomena of field line stretching (leading to field amplification) and flux expulsion. From a dynamical point of view, we have the possibility of wave like phenomena (Alfvén and magnetoacoustic waves). In addition, we have the effects of the amplified fields on the flows, which can lead to non-trivial scale-selection effects when strong fields interact with thermal convection, as in the solar photosphere. This chapter aims to give a brief discussion of these areas.

2 Basic Processes

2.1 Kinematics

The influence of a velocity field $\mathbf{u}$ on a solenoidal magnetic field $\mathbf{B}$ is described (ignoring relativistic effects) by the induction equation

$$\frac{\partial \mathbf{B}}{\partial t} = \nabla \times (\mathbf{u} \times \mathbf{B}) - \nabla \times (\eta \nabla \times \mathbf{B}) \tag{1}$$

where the magnetic diffusivity η may depend on position (as it is a function of temperature). Here we have implicitly assumed an isotropic Ohm's Law, and neglected such features of plasma charge transport as the Hall effect, though this may be significant near the very top of the solar atmosphere.

Away from sunspots and other magnetic structures the solar convection zone presents a cellular appearance on many scales (Figure 1). The smallest scales of motion are the *granules* which have horizontal scales of from hundreds of some thousands of kilometres and lifetimes of c. 15 min. There is strong evidence of a larger ordering of the cells (the *supergranulation*), associated with motion at greater depth in the convection zone. Observations of the magnetic field show that there are regions of strong field in the interstices of the granular and supergranular cellular structure, which are precisely the regions of converging horizontal flow. We can account for this by the effects of Faraday induction at large values of R_m.

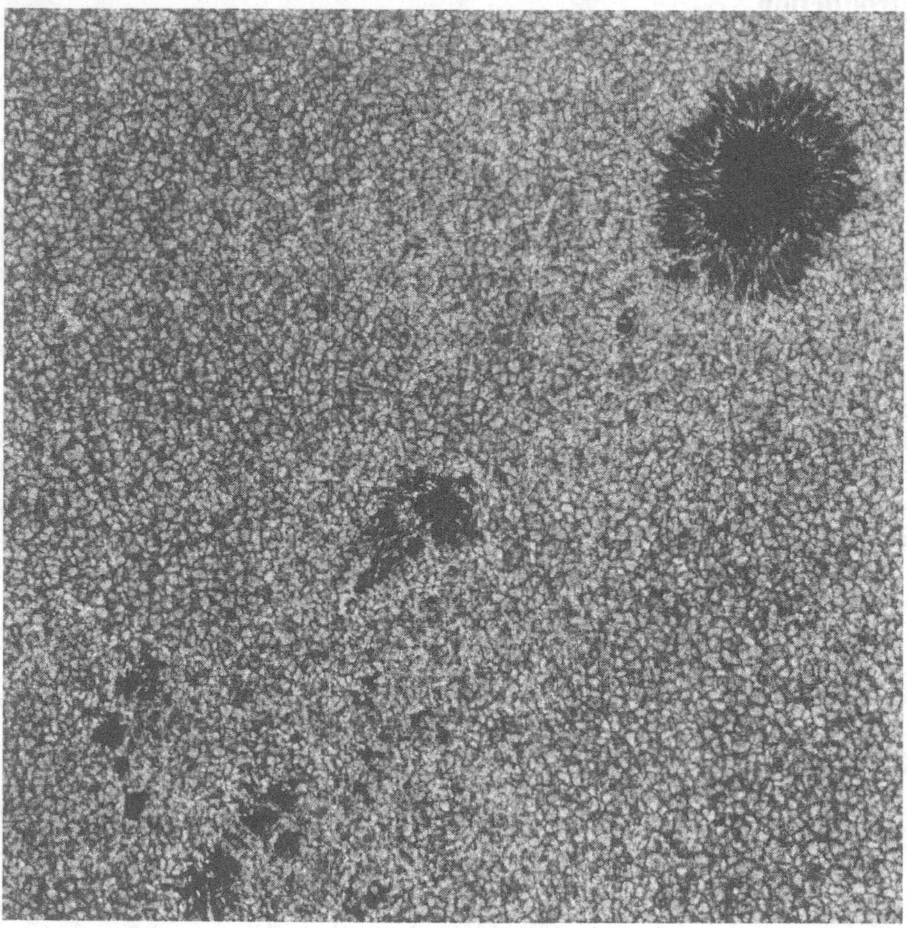

Figure 1. View of part of the solar photosphere in visible light (courtesy of Dr T.Berger). Shown are the normal granulation, a large sunspot (top right), smaller pores; notice the abnormally narrow cells near the magnetic features

There is no doubt that if we adopt molecular values for η of the order $10^2 - 10^3\ \mathrm{m^2s^{-1}}$ in the upper regions of the convection zone, then the diffusivity is only important (for the time scales appropriate to convective evolution) on very short length scales of the order of 1 – 10 km. When diffusivity can be neglected, (2.1) leads to *Alfvén's theorem* that magnetic field lines move with the fluid. Thus one might expect magnetic fields to be concentrated in regions of converging flow, as observations suggest. The final distribution of magnetic field in a prescribed flow will depend on diffusion, as numerical computations in simple geometries show. Figure 2, from Galloway & Proctor (1983), shows that for a steady flow with hexagonal symmetry (crudely modelling conditions in a granule) the initial tendency is for all vertical flux to be concentrated in converging flow regions, while in the final, diffusively limited steady state only about half the flux remains near points of flow convergence.

The other important property of large R_m flows is *field line stretching*; here the direct analogue is with vortex stretching in hydrodynamics; tubes of flux which increase in length decrease in diameter, since the flux is conserved the field strength rises, and integrating over all tubes the total magnetic energy is increased.

2.2 The Lorentz Force

The magnetic field reacts with the flow in two ways: through an ohmic heating term in the energy balance, which does not seem to be of much importance in most applications, and through a body force acting directly on the fluid. This (Lorentz) force has the form

$$\mathbf{J} \times \mathbf{B} = \frac{1}{\mu_0}(\nabla \times \mathbf{B}) \times \mathbf{B} = \frac{1}{\mu_0}(\mathbf{B} \cdot \nabla \mathbf{B} - \nabla(\frac{1}{2}|\mathbf{B}|^2)) \tag{2}$$

The term $\mathbf{B} \cdot \nabla \mathbf{B}$ on the r.h.s. of (2.2) is only effective when the field lines are curved, and takes the form of a restoring force that tends to straighten the lines. The other, gradient term is the *magnetic pressure* and demonstrates the tendency of a concentration of field to expand. In the large R_m regime, even when the ambient (average) field is weak, the effects of the flow on the field can be such that the field lines are significantly curved and the local field strength is large. Both these parts of the Lorentz force inhibit flux concentration, the former by reducing local fluid velocities and the latter by reducing the local gas pressure and inducing a buoyancy force that acts against the converging flow. This latter property leads to the phenomenon of *magnetic buoyancy* or *Parker instability* in which regions of strong horizontal field, being buoyant, will rise under gravity. Such strong fields are thought to be produced near the bottom of the solar convection zone, at 70% of the solar radius. The effects of the buoyancy are presumably to permit great loops of strong field to rise to the surface, though the mechanism is not wholly understood. For some details see Wissink et al. (2000). It is likely that the vast majority of flux that arises at the surface is due to these buoyant loops. In addition, the restoring (curvature) forces allow the system to support waves if diffusivities are sufficiently low; thus in a superadiabatically stratified layer new forms of convection, involving oscillatory motion, can occur.

2.3 Alfvén Waves

There has been some discussion of Alfvén waves in other chapters for the case of incompressible flow. In the Solar convection zone the fluid is highly compressible and the situation is more

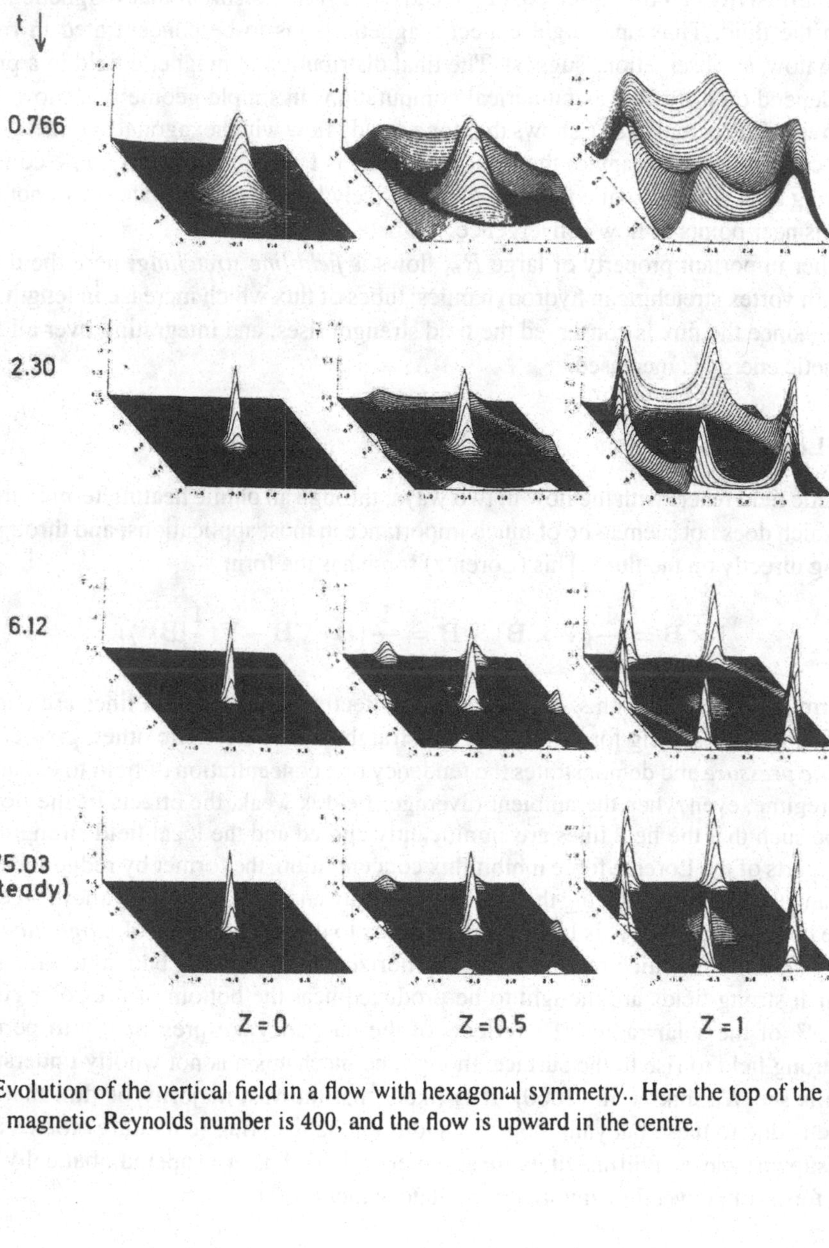

Figure 2. Evolution of the vertical field in a flow with hexagonal symmetry.. Here the top of the layer is at $z = 1$. The magnetic Reynolds number is 400, and the flow is upward in the centre.

complicated. We ignore diffusion in Eq. (2.1) and also consider the momentum and continuity equations. Linearizing about a basic static state in which $\mathbf{B} = \mathbf{B}_0$ (uniform) and density $\rho = \rho_0$, the equations for the velocity $\mathbf{u}$, perturbation density ρ and perturbation field $\mathbf{b}$ are given by (writing $\mathbf{V}_0 \equiv (\mu_0\rho_0)^{-\frac{1}{2}}\mathbf{B}_0$), $\mathbf{v} \equiv (\mu_0\rho_0)^{-\frac{1}{2}}\mathbf{b}$ where $\mathbf{V}_0$, $\mathbf{v}$ are the respective *Alfvén speeds*, we find

$$\dot{\mathbf{u}} = -c^2\nabla\rho/\rho_0 + \mathbf{V}_0\cdot\nabla\mathbf{v} - \nabla(\mathbf{V}_0\cdot\mathbf{v}) \tag{3}$$

$$\dot{\mathbf{v}} = \mathbf{V}_0\cdot\nabla\mathbf{u} - \mathbf{V}_0\nabla\cdot\mathbf{u} \tag{4}$$

$$\dot{\rho} = -\rho_0\nabla\cdot\mathbf{u} \tag{5}$$

If we seek solutions proportional to $e^{i\mathbf{k}\cdot\mathbf{x}-i\omega t}$ then we obtain the dispersion relation, where c_A is the Alfvén speed $|\mathbf{V}_0|$,

$$(\omega^2 - (\mathbf{V}_0\cdot\mathbf{k})^2)(\omega^4 - \omega^2k^2(c^2 + c_A^2) + c^2k^2(\mathbf{V}_0\cdot\mathbf{k})^2) = 0 \tag{6}$$

The first bracket in this relation shows that there are waves that travel at the Alfvén velocity $\mathbf{V}_0$ even in a compressible atmosphere. These are the true Alfvén waves, and they are characterized by being *torsional*, having no field or flow parallel to $\mathbf{V}_0$ (in fact the vorticity and perturbation current are $\parallel \mathbf{V}_0$). Clearly such motions do not compress the gas, so c is not expected to appear. The remaining bracket can be further factorized into two other types of wave, called *fast* and *slow magnetoacoustic waves* according to their phase speed. They are characterized as follows:

(i) $c^2 \gg c_A^2$; then fast waves are almost all longitudinal (compressional) with speeds $\approx c$ are are almost isotropic while slow waves are almost transverse with speeds $\lessapprox c_A$ and travel along the field lines.

(ii) $c^2 \ll c_A^2$; then slow waves are highly anisotropic, taking the form of 1-dimensional sound waves with flow parallel to $\mathbf{V}_0$, while fast waves are almost transverse.

Except quite near the surface of the Sun case (i) applies, and in this case the slow waves are often called Alfvén waves. It is these waves that are destabilized by convection.

3 Convection in Sunspots

3.1 Observations

Not all the solar surface is occupied by ordinary granulation. There are regions of highly concentrated field – sunspots and pores. These usually emerge from below the surface in a *bipolar region*, showing that they are the manifestations of large flux tubes that erupt from the large scale zonal field near the bottom of the convection zone, as discussed above. Figure 1 shows a large sunspot, with surrounding granular flows. Note the dark centre (umbra) of the spot, surrounded by the filamentary penumbra. In the umbra very strong fields, up to 0.2-0.4 tesla, act to provide a strong restoring force on the motion and impede the amplitude of convection. Since energy must still be removed from the solar surface, the radiative flux must increase in the neighbourhood of a spot, which is why the spot appears cool and dark. Convection is still going on in the umbra, however, and can be observed in the form of "umbral dots", occasional bright points in the darkness of the umbra. In the penumbra the field is still quite strong, but is no longer predominantly vertical. An excellent discussion of modern observations is given in Title (2000). In fact the field is nearly horizontal near the outside of the umbra, and the Lorentz forces due to this field favour motion that varies little along the field lines (so as to reduce the bending and consequent curvature forces). Thus the picture there is of quasi-two-dimensional rolls, with light and dark elements representing hot (rising) and cold (falling) plasma. Near the boundary of the sunspot, there are granular cells, but with a different morphology from those far from the spot. Although the relatively smooth appearance of the spot may be a surface phenomenon only, with theorists disagreeing on whether the spot is coherent or filamentary at greater depth, it is certainly true that some aspects of sunspot dynamics can be understood in terms of simple convection models. An excellent introduction to sunspots is given in Thomas & Weiss (1992) We give details only for the simplest situation, but illustrate by reference to more realistic models.

3.2 Boussinesq magnetoconvection

Consider a layer of fluid between horizontal boundaries, and a uniform imposed vertical magnetic field. The layer is heated from below. We assume that the boundaries, at $z = 0, d$, are kept at the fixed temperatures $T_0 + \Delta T, T_0$. We suppose that the boundaries are *stress-free*, *impermeable* with the magnetic field constrained to be *vertical*. There is an imposed magnetic field $\mathbf{B}_0\hat{\mathbf{z}}$. We shall assume the Boussinesq approximation, so that the fluid is (effectively) incompressible. This problem has been addressed in detail by Proctor & Weiss (1982). Then the equations are

$$\frac{1}{\sigma}\left(\frac{\partial \mathbf{u}}{\partial t} + \mathbf{u}\cdot\nabla\mathbf{u}\right) = -\nabla p + R\theta\hat{\mathbf{z}} + Q\zeta\left[\frac{\partial \mathbf{b}}{\partial z} + \mathbf{b}\cdot\nabla\mathbf{b}\right] + \nabla^2\mathbf{u} \qquad (7)$$

$$\frac{\partial\theta}{\partial t} + \mathbf{u}\cdot\nabla\theta = \mathbf{u}\cdot\hat{\mathbf{z}} + \nabla^2\theta \qquad (8)$$

$$\frac{\partial \mathbf{b}}{\partial t} = \nabla\times(\mathbf{u}\times\mathbf{b}) + \frac{\partial \mathbf{u}}{\partial z} + \zeta\nabla^2\mathbf{b} \qquad (9)$$

$$\nabla\cdot\mathbf{b} = \nabla\cdot\mathbf{u} = 0 \qquad (10)$$

where the temperature $T = T_0 + \Delta T(1 - \frac{z}{d} + \theta)$, the magnetic field $\mathbf{B} = B_0(\hat{\mathbf{z}} + \mathbf{b})$, the velocity is scaled with κ/d, where k is the thermal diffusivity and time is scaled with d^2/κ.

The dimensionless parameters are: the Prandtl number $\sigma = \nu/\kappa$; ν is the (kinematic) viscosity. The diffusivity ratio $\zeta = \eta/\kappa$ where $\eta = 1/\mu_0\sigma$ is the magnetic diffusivity. The Rayleigh number $R = \frac{g\alpha\Delta T d^3}{\kappa\nu}$ where α is the coefficient of expansion, and the Chandrasekhar number $Q = (B_0^2 d^2)(\mu_0\rho\kappa\eta)^{-1}$ (proportional to (Hartmann number^2)).

Note that in the laboratory, $\zeta \gg 1$, while in astrophysical situations ζ can be 0(1) or even smaller, if κ is enhanced by effects of radiation, eg in stellar atmospheres.

These equations are to be solved with boundary conditions at $z = 0, 1$ (dimensionless)

$$\mathbf{u}\cdot\hat{\mathbf{z}} = 0,\ \frac{\partial}{\partial z}(\mathbf{u}\times\hat{\mathbf{z}}) = 0,\ \theta = 0,\ \mathbf{b}\times\hat{\mathbf{z}} = 0\,. \qquad (11)$$

If we linearize, equations are separable in x, y, t. So can look for solutions with e.g. $\theta = \hat{\theta}(z)e^{st+i\mathbf{k}\cdot\mathbf{x}}$, $\mathbf{k} = (k_x, k_y, 0)$, $k = |\mathbf{k}|$, etc.

Inspection of the equations shows that solutions can be found in the form

$\mathbf{b} = (\tilde{b}_x \sin \pi z, \tilde{b}_y \sin \pi z, \tilde{b}_z \cos \pi z)$,

$\mathbf{u} = (\tilde{u}_x \cos \pi z,\ \tilde{u}_y \cos \pi z,\ \tilde{u}_z \sin \pi z)$,

$\hat{\theta} = \tilde{\theta} \sin \pi z$.

It may also be deduced from the equation that

$$\frac{1}{\sigma}\dot{\omega}_z = Q\zeta\frac{\partial j_z}{\partial z} + \nabla^2\omega_z \tag{12}$$

$$\dot{j}_z = \frac{\partial \omega_z}{\partial z} + \zeta\nabla^2 j_z \tag{13}$$

where

$$\omega_z = (\nabla \times \mathbf{u})_z,\ j_z = (\nabla \times \mathbf{b})_z\ . \tag{14}$$

Thus these variables only admit decaying solutions (these are in fact just the torsional Alfvén waves described in an earlier lecture), and we are left only velocities and fields supported by horizontal vorticity and current. Thus eg $\hat{\mathbf{u}} = (+i\pi k_x\tilde{\phi}\cos\pi z,\ i\pi k_y\tilde{\phi}\cos\pi z,\ k^2\tilde{\phi}\sin\pi z)$, with a similar expression for $\hat{\mathbf{z}}$. Then substituting into the equations and eliminating $\tilde{\theta}$, $\tilde{\phi}$ etc. we arrive at the dispersion relation

$$\beta^2(s+\beta^2)(s+\sigma\beta^2)(s+\zeta Q\beta^2) + \sigma\delta\theta\beta^2\pi^2(s+\beta^2) - R\sigma k^2(s+\zeta\beta^2) = 0 \tag{15}$$

where $\beta^2 = k^2 + \pi^2$.

Note that this "simple" relation depends on the boundary conditions adopted.

We consider first steady solutions ($s = 0$). Then we have

$$R = \frac{\beta^6 + Q\pi^2\beta^2}{k^2} \equiv R_{(e)}$$

which reduces to the usual condition $R = \beta^6/k^2$ for $Q = 0$.

When $Q = 0$ (no magnetic field) the marginal curve of $R_{(e)}$ against k has a single minimum of $27\pi^4/4$ when $k = \pi/\sqrt{2}$ For small Q the preferred wavenumber (where $R_{(e)}$ is least) is still O(1), while for large Q we find that

$$R \sim \pi^2 Q,\ k = O(Q^{\frac{1}{6}}) \gg 1\ .$$

Thus we see that the effect of the strong magnetic field is to produce tall, thin convection cells aligned with the imposed field.

More interesting is the prospect of oscillatory convection. It must be emphasised that such behaviour can only occur when the conductivity of the plasma is sufficiently high, and relies crucially on finite magnetic Reynolds number (which must be larger than the Peclet number). At onset the growth rate $s = i\omega$, where ω is real. Substituting into the equation and taking real and imaginary parts gives two relations involving ω and R. Solving these gives

$$R = R_{(0)} = \frac{\beta^6}{k^2}\left[1 + \frac{\zeta}{\sigma}(1+\sigma+\zeta)\right] + \frac{\pi\beta^2 Q}{k^2}\left[\frac{(\sigma+\zeta)\zeta}{1+\sigma}\right] \tag{16}$$

$$\omega^2 = \frac{k^2}{\beta^2}\left(\frac{\sigma\zeta}{1+\sigma+\zeta}\left(R_{(e)} - R_{(0)}\right)\right) = \beta^4\left[-\zeta^2 + \frac{(1-\zeta)\zeta\sigma}{1+\sigma}\frac{\pi^2 Q}{\beta^4}\right]\ . \tag{17}$$

Note that, as in the steady case, the critical value $R_{(0)}$ only depends on $k = |\mathbf{k}|$ and not on the direction of $\mathbf{k}$. Note also that $\omega^2 > 0$ only if $\zeta < 1$, and if Q is sufficiently large. The first form of the expression for ω^2 shows that oscillations are possible only when they arise at a lower value of R than steady-state convection.

These overstable oscillations are really Alfvén waves, destabilized against Ohmic and viscous decay by the unstable temperature gradient. They transport very little heat, and give way to overturning convection when R is sufficiently large.

3.3 Nonlinear theory

To look at these finite amplitude effects, we need nonlinear theory. Of course, now the solutions are of finite amplitude and the vigour of the convection is a function of R, Q and the other parameters. In addition, there is the important problem of planform selection (see below). The instability criteria do not distinguish between different horizontal forms of motion, since any sum $\sum \alpha_j e^{i\mathbf{k}_j\cdot\mathbf{x}}$ is a solution provided all the $|\mathbf{k}_j|$ are the same.

We first investigate nonlinear effects. This has been done thoroughly only for two-dimensional motion, where $u_y = b_y \equiv 0$, and where there is no y dependence. Then numerical methods may be used to investigate possible forms of motion.

Choosing periodic boundary conditions in x (and y), we can fix the spatial period λ. Then we get several possibilities for nonlinear steady convection. Plotting the energy E of the motion (which can be defined in a number of ways) against R, with the other parameters fixed, we find three characteristic types of behaviour for steady motion, as shown in Figure 3; for further discussion see Proctor & Weiss (1982). Thus the bifurcation at $R = R_{(0)}$ can be either supercritical

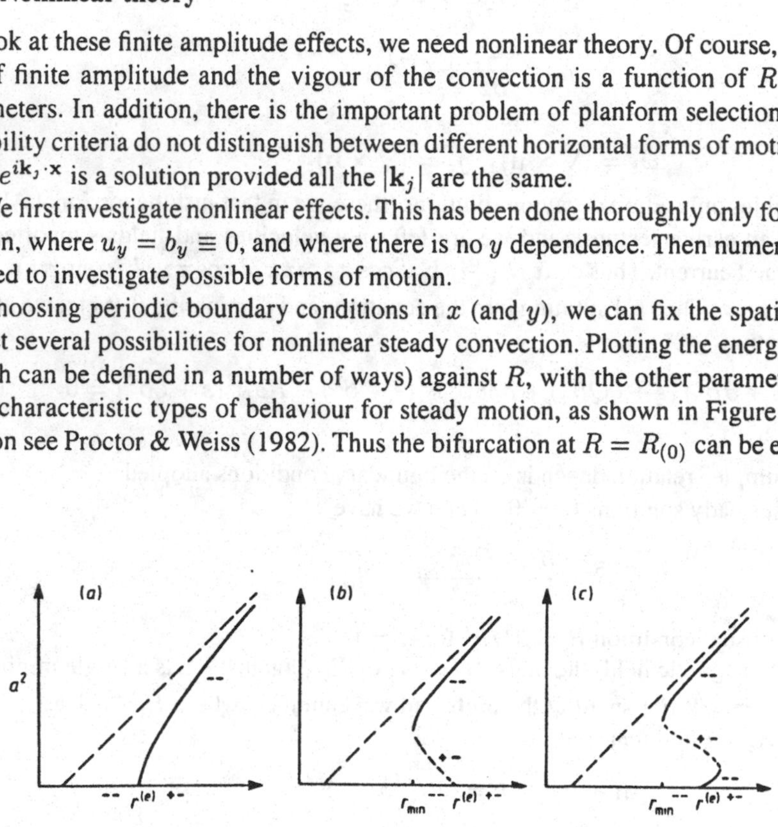

Figure 3. Dependence of amplitude of motion on R in three cases of nonlinear two-dimensional magneto-convection

or subcritical depending on the parameters. The reason for subcriticality can be understood by recalling the effects of flux expulsion discussed earlier.

Consider steady flows, then the induction equation is (writing $\mathbf{b} + \hat{\mathbf{z}} = \mathbf{B}$);

$$0 = \nabla \times (\mathbf{u} \times \mathbf{B}) + \zeta \nabla^2 \mathbf{B}$$

In two dimensions, we can write $\mathbf{B} = \nabla \times (A\hat{\mathbf{y}})$ and then the equation can be written (uncurling and taking the y component)

$$0 = -\mathbf{u} \cdot \nabla A + \zeta \nabla^2 A$$

Now if $|\mathbf{u}|R_m/\zeta \gg 1$ then at leading order $\mathbf{u} \cdot \nabla A \approx 0$ or A is constant on streamlines at least away from boundaries. However integrating round a closed streamline (they are essentially all closed in the 2-D case) we get the exact result

$$\oint \frac{1}{|\mathbf{u}|} \mathbf{u} \cdot \nabla A d\mathbf{x} = 0 = \oint \frac{1}{|\mathbf{u}|} \zeta \nabla^2 A d\mathbf{x} \tag{18}$$

which implies that A does not vary across streamlines either. Thus A is uniform away from boundaries and magnetic field is confined to boundary layers. To estimate their thickness δ we balance diffusion, $O(A\zeta/\delta^2)$ with advection $(\mathbf{u} \cdot \nabla A)$, near a stagnation line where $\mathbf{u} = O(\delta)$, so $\mathbf{u} \cdot \nabla$ is independent of δ. Thus $\delta \propto \zeta^{1/2}$ or $\delta = O(R_m^{-1/2})$.

Since the magnetic field is only at the edges of the cells, in wide cells it interferes less with the motion than at low amplitudes. We can construct an energy integral by taking products of the equations with $\mathbf{u} \cdot \hat{\mathbf{z}}$ and $\mathbf{B}$. In the steady state we have the relation (with $\langle \cdot \rangle$ representing an integral over the periodic domain)

$$R\langle|\nabla\theta|^2\rangle = \langle|\nabla\mathbf{u}|^2\rangle + Q\zeta^2\langle|\nabla\mathbf{B}|^2\rangle \tag{19}$$

Now when $|\nabla\mathbf{B}|$ is small, (low R_m, $\mathbf{B} = O(\frac{1}{\zeta})$), $\mathbf{B}$ varies on an O(1) length scale and so the stabilizing magnetic term is independent of ζ. When $\mathbf{B}$ is confined to thin flux sheets, however, with thickness $O(\zeta^{1/2})$, we have peak fields of order $\zeta^{-1/2}$ (since the total flux $\int_0^\lambda B_z dx$ is a constant). Thus $|\nabla\mathbf{B}| = O(\zeta^{-1})$ and so $\langle|\nabla\mathbf{B}|^2\rangle \sim \zeta^{-3/2}$. Hence the stabilizing term $Q\zeta^2\langle|\nabla\mathbf{B}|^2\rangle = 0(\zeta^{1/2})$ which is small for small ζ.

Thus convection can exist for $R \ll R_{(c)}$ for some values of the parameters. How do the oscillatory motions fit in when overstability is possible? Some typical scenarios are shown in Figure 4, from Proctor & Weiss (1982). There are two generic possibilities. When the steady branch is supercritical, the oscillatory branch becomes vacillatory (small wiggles about an (unstable) steady solution) and these wiggles disappear in an oscillatory bifurcation, after which the steady solution is stable. When the steady branch is subcritical, the period of the oscillations becomes longer and they finally disappear in a heteroclinic bifurcation on the unstable steady branch. The system then jumps abruptly to a larger amplitude steady solution.

4 Planform selection

An important question in the nonlinear theory of convection is what planform of motion is realised near onset. The answer to this question depends on the periodic domain chosen. Let us choose for our domain a square box (size λ) in the x and y directions. Then we choose the box size so that $\lambda = 2\pi/k_c$ where k_c is the optimum wave number from the linear theory. (These optima are different for steady and oscillatory motions.) Then linear theory admits solutions proportional to $e^{ik_c x}$ and $e^{ik_c y}$, if for example $w \propto e^{ik_c x}$ the solution is called

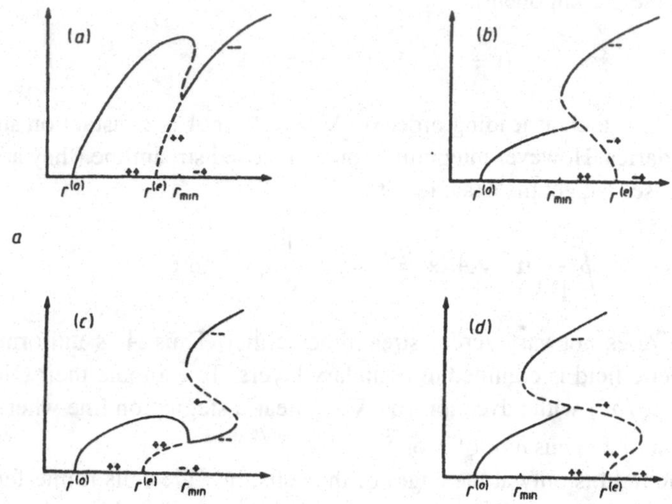

Figure 4. Sketches of the behaviour of the amplitude of steady and oscillatory convection as a function of R for various parameter values.

rolls while if $w \propto e^{ik_c x} + e^{ik_c y}$ the solution is called squares. Since both these solutions have the same horizontal wave number k_c they are both equally good. But they differ in their nonlinear characteristics; and by performing perturbation theory we can distinguish between them. The case of steady motion is easiest. We let $R = R_{(e)} + \epsilon^2 R^2$ and write e.g. $w = \epsilon R_{(e)} \left[(Ae^{ik_c x} + Be^{ik_c y}) \right] \sin \pi z + \epsilon^2 w_2 + \epsilon^3 w_3 + \ldots$ where A and B are (complex) functions of time. By expanding the equations in powers of ϵ, and using perturbation theory, we can derive coupled equations for A and B. They may after scaling be reduced to the form

$$\frac{dA}{dt} = \mu A - a|A|^2 A - b|B|^2 A + \ldots \tag{20}$$

$$\frac{dB}{dt} = \mu B - a|B|^2 B - b|A|^2 B + \ldots \tag{21}$$

where A and B are real coefficients depending on the parameters, and μ (real) is proportional to $R - R_{(e)}$. The form of the equations is fixed by symmetry considerations. They must be unchanged under the following operations:

$$\text{Reflection in } x-\text{axis} \qquad A \to A^*, \quad B \to B \tag{22}$$

$$\text{Reflection in } x = y \qquad A \to B, \quad B \to A \tag{23}$$

$$\text{Translations in } x \text{ and } y \qquad A \to Ae^{i\delta_1}, \; B \to Be^{i\delta_2} \tag{24}$$

We then find that the steady solutions (rolls) $B = 0, |A|^2 = \mu/a$ [or $A = 0, |B|^2 = \mu/a$] is always stable, while the other solution $|B|^2 = |A|^2 = \mu/(a+b)$ (squares) is unstable. So for Boussinesq magnetoconvection rolls are the stable form of steady convection.

The situation for oscillations is more complex. Now there are four different types of basic solution, each representing motions in the form of travelling waves. So

$$w = Re\left[A_1 e^{ik_c x+i\omega t} + A_2 e^{-ik_c x+i\omega t} + B_1 e^{ik_c y+i\omega t} + B_2 e^{-ik_c y+i\omega t}\right] \tag{25}$$

Thus we will have four complex coupled equations to solve!

The actions of the symmetries in this case are:

$$\text{Reflection in} x = 0: \quad A_1 \to A_2 \quad A_2 \to A_1 \quad B_i \to B_i \tag{26}$$

$$\text{Reflection in} x = y: \quad A_1 \leftrightarrow B_1 \quad A_2 \leftrightarrow B_2 \tag{27}$$

$$\text{Translation in } x, y \text{ and time}: \quad A_1 \to A_1 e^{i\delta_1} \quad A_2 \to A_2 e^{-i\delta_1} \quad B_i \to B_i; \tag{28}$$

$$A_i \to A_i \quad B_1 \to B_1 e^{i\delta_2} \quad B_2 \to B_2 e^{-i\delta_2}; \tag{29}$$

$$A_i \to A_i e^{i\delta} \quad B_i \to B_i e^{i\delta} \tag{30}$$

Then the equations are

$$\dot{A}_1 = (\mu + i\omega)A_1 + A_1\left(a|A_1|^2 + b|A_2|^2 + c(|B_1|^2 + |B_2|^2)\right) + eA_2^* B_1 B_2 \tag{31}$$

where a, b, c and e are complex, together with three other similar equations.

There are now five types of solution at onset:

$$\text{Travelling rolls} \quad (A_1 \neq 0, A_2 = B_1 = B_2 = 0) \tag{32}$$

$$\text{Travelling squares} \quad (A_1 = B_1 \neq 0, A_2, B_2 = 0) \tag{33}$$

$$\text{Standing rolls} \quad (A_1 = A_2 \neq 0, B_1, B_2 = 0) \tag{34}$$

$$\text{Standing squares} \quad (A_1 = A_2 = B_1 = B_2 \neq 0) \tag{35}$$

$$\text{Alternating rolls} \quad (A_1 = A_2 = iB_1 = iB_2) \tag{36}$$

All these types can be stable depending on the parameters (see the diagram in Clune & Knobloch, 1994). The predominant stable form of oscillation is the Alternating Roll pattern, as shown in Figure 5, consisting of two orthogonal oscillating roll patterns, 90 degrees out of phase. These patterns in turn can become unstable to secondary instabilities leading to a wide variety of less symmetric solutions. Details may be found in Rucklidge et al. (2000).

At higher R the regular patterns break down and we get disordered flow. Of particular current interest is the coexistence of two different types of flow (a) large field, small velocity; (b) field expelled, large velocity. This is known as *flux separation*. The numerically computed solutions bear a strong resemblance to the observed behaviour of solar magnetic fields. Figure 1 shows a view of the solar granulation near an active region. Visible are the normal convection cells, anomalous narrow convection near the dark magnetic features, a sunspot, with umbra and penumbra, and smaller 'pores' without a penumbra. Numerical computations have not yet been able to model the life of a sunspot in detail, recent three-dimensional simulations in Cambridge show some of the effects of a magnetic field at large R_m on turbulent convection in a gas. For these simulations, instead of Boussinesq convection, we took a layer of perfect gas, between two layers. The action of gravity sets up a hydrostatic pressure gradient, and thus a density gradient in the gas. In the computations shown here, the density contrast in the basic state is a factor of 11, so that

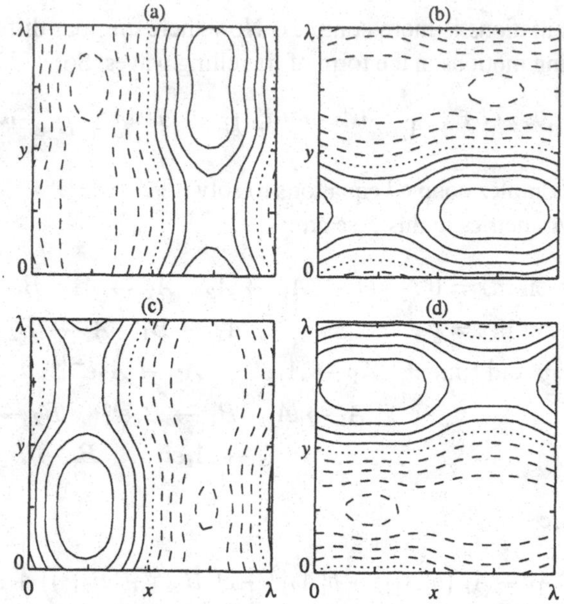

Figure 5. Four stages in the alternating roll solution for marginal convection. The contours represent values of e.g. temperature or vertical velocity.

compressibility is very important. The equations of motion and heat transfer must be supplemented by the perfect gas equation and the continuity equation. Further details may be found in e.g. Weiss et al. (1990), Weiss et al. (1996). Figure 6 shows snapshots of the evolved statistically steady states for three different values of Q. It is evident that two scales of motion are simultaneously possible, and the proportion of field-free, large scale motion increases with decreasing magnetic field strength. For large Q the convection is dominated by cells with small horizontal extent and is fairly well ordered. For small Q, on the other hand, the horizontal extent of the cells is large, the convection is disordered and aperiodic, and the magnetic field, which is now dynamically passive, is pushed around in the interstices of the cells. Interestingly, this aperiodic response leads to sheets of flux rather than tubes, when the field is weak. For stronger fields, the field becomes more tubelike.

5 Conclusion

In this short chapter I have tried to give a brief overview of some of the principal ideas behind MHD at high magnetic Reynolds number. The chapter does not cover every topic and concentrates mainly on convection. An important subject not addressed here is that of magnetostatic equilibria and force-free fields, whose study is important in the understanding of the solar corona, and features such as prominences. In the area of MHD convection, the lack of any possible terrestrial experimental verification of theoretical and numerical predictions has been a problem in the past. However more and more detailed observations of magnetic structures on the Sun has for

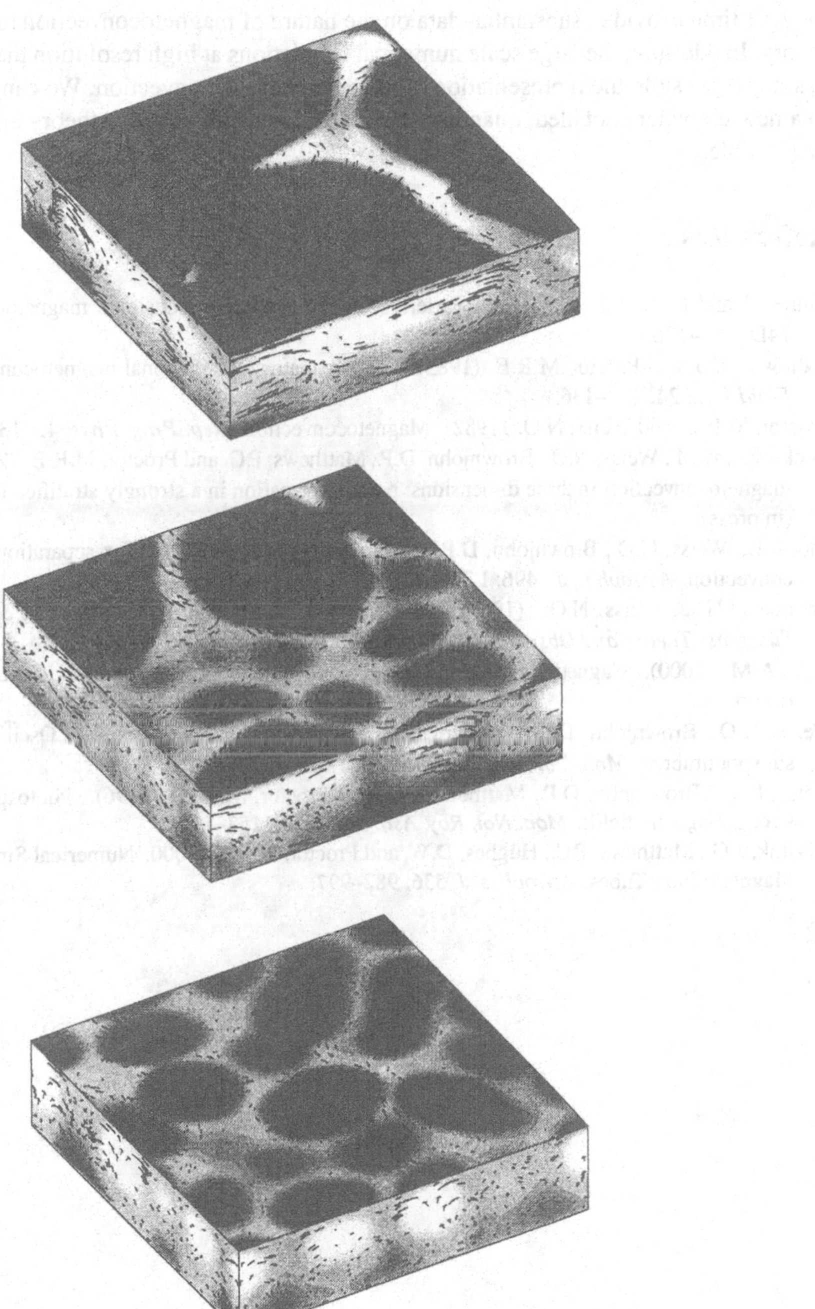

Figure 6. Flux separation. Three different snapshots of the nature of teh magnetic field at the top of the layer (lighter grey denotes stronger field). Q increases downward. Note the two different forms of convection coexisting in the middle picture.

the first time provided substantial data on the nature of magnetoconvection to compare with the theory. In addition, the large scale numerical simulations at high resolution that are now possible has made possible the representation of the details of the convection. We can now look forward to a new era where detailed, quantitative comparisons with between theory and observation will be possible.

References

Clune, T. and Knobloch, E. (1994). Pattern selection in three-dimensional magnetoconvection. *Physica* 74D: 151–176.

Galloway, D.J. and Proctor, M.R.E. (1983). The kinematics of hexagonal magnetoconvection. *Geo. Astro. Fluid Dyn.* 24: 109–136.

Proctor, M.R.E. and Weiss, N.O. (1982). Magnetoconvection. *Rep. Prog. Phys.* 45: 1317–1379.

Rucklidge, A.M., Weiss, N.O., Brownjohn, D.P., Matthews, P.C. and Proctor, M.R.E. (2000). Compressible magnetoconvection in three dimensions: pattern formation in a strongly stratified layer. *J.Fluid Mech.*: (in press).

Tao, L.L., Weiss, N.O., Brownjohn, D.P.,and Proctor, M.R.E. (1998). Flux separation in stellar magnetoconvection. *Astrophys. J.*. 496: L39–L42.

Thomas, J.H. & Weiss, N.O. (1992). The theory of sunspots. In Thomas, J.H & Weiss, N.O., eds., *Sunspots: Theory and Observations.* Dordrecht: Kluwer.

Title, A.M. (2000). Magnetic fields below, on and above the solar surface. *Phil. Trans Roy. Soc. Lond. A*: (in press).

Weiss, N.O., Brownjohn, D.P., Hurlburt, N.E. and Proctor, M.R.E. (1990). Oscillatory convection in sunspot umbrae. *Mon. Not. Roy. Astr. Soc.* 245: 434–452.

Weiss, N.O., Brownjohn, D.P., Matthews, P.C. and Proctor, M.R.E. (1996). Photospheric convection in strong magnetic fields. *Mon. Not. Roy. Astr. Soc.* 283: 1153–1164.

Wissink, J.G., Matthews, P.C., Hughes, D.W. and Proctor, M.R.E. 2000. Numerical Simulation of Buoyant Magnetic Flux Tubes. *Astrophys. J.* 536: 982–997.

Chapter IV

A. Thess, D. Schulze

COMPUTATIONAL MAGNETOHYDRODYNAMICS PART I – FUNDAMENTALS

Computational Magnetohydrodynamics
Part I - Fundamentals

André Thess[1], Dietmar Schulze[2]

[1]Department of Mechanical Engineering, Ilmenau University of Technology,
P.O. Box 10 05 65, 986984 Ilmenau, Germany,
Electronic address: thess@tu-ilmenau.de
[2]Department of Electrical Engineering, Ilmenau University of Technology,
P.O. Box 10 05 65, 986984 Ilmenau, Germany,
Electronic address: dietmar.schulze@e-technik.tu-ilmenau.de

1. Introduction

Computational Magnetohydrodynamics (CMHD) is the science of numerically solving the coupled set of equations of fluid dynamics and electrodynamics occasionally supplemented by mathematical models for phase transitions including solidification, melting, evaporation and condensation.

Figure 1 shows the levitation melting process as an illustrative example of a problem which presents a fundamental and still unsurmounted challenge to CMHD. An electrically conducting solid body (e. g. a sphere as in Fig. 1a) is placed above a coil which carries an alternating current $I_0 \cos(\omega t)$. The resulting magnetic filed penetrates the body inducing eddy currents **J**. These currents produce a heat flux $\dot{q} = \mathbf{J}^2/\sigma$ (where σ denotes the electrical conductivity) which leads to the melting of the metal (Fig. 1b). At the same time, a Lorentz force $\mathbf{F} = \mathbf{J} \times \mathbf{B}$ is created within the body which may support the solid or molten body against gravity, and which induces a sometimes vigorous turbulent motion inside the fluid.

In order to apply levitation melting or the closely related cold crucible technique (Asai 2000 and references therein) to the containerless processing of reactive materials or measurement of thermophysical properties (Egry & Szekely 1991) one has to answer the following question:

"Which values of electric current I_0, frequency ω, and what shape of the coil are best suited to rapidly melt and stably levitate a given quantity of metal?"

It becomes apparent by inspection of Fig. 1c that this computational task is neither a pure fluid dynamic nor a pure electrodynamic problem. In order to compute the magnetic field which, for sufficiently high frequency, penetrates only into a skin layer with thickness $\delta = (2/\mu_0\sigma\omega)^{1/2}$, we need in general to solve the equations of free-surface hydrodynamics under the influence of the unknown electromagnetic force and heat flux. As of today (2000), a numerical simulation of this conceptually simple problem including electromagnetics, free-surface deformation, and fully resolved turbulent flow inside the liquid metal is beyond the capacity of even the most powerful computers.

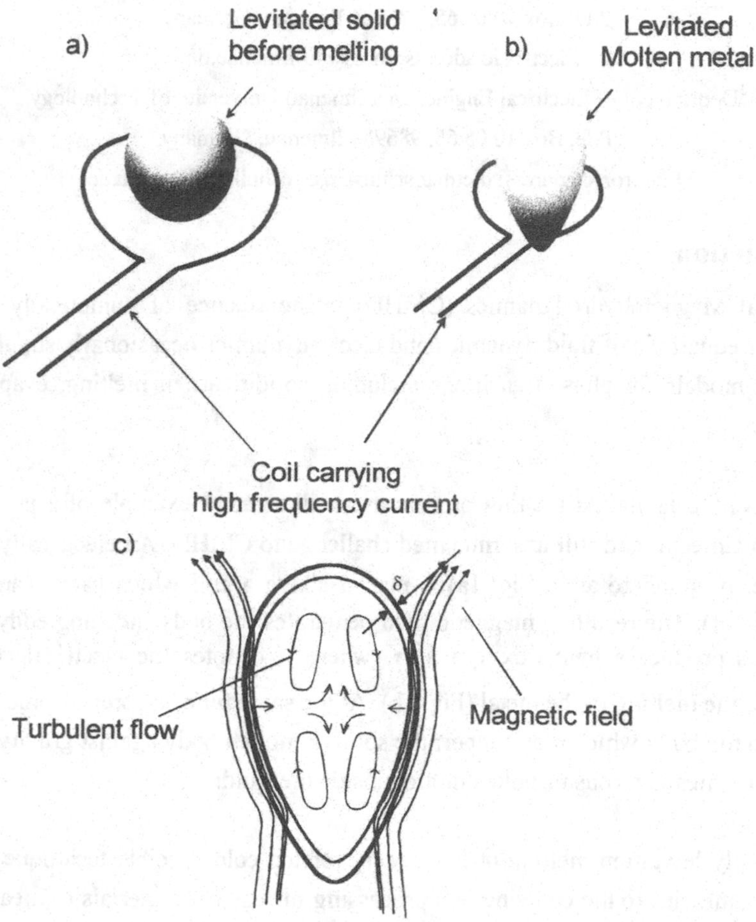

Figure 1: Illustration of the levitation-melting process (after Sneyd & Moffatt 1982) (a) and (b) idealized levitation device before and after melting, (c) sketch of the spatial structure of magnetic field and velocity field.

The example of electromagnetic levitation melting illustrates the subtle closed-loop interaction between electromagnetics and fluid dynamics in CMHD which is characteristic of a wide variety of magnetohydrodynamic problems. The scope of CMHD ranges from fundamental studies of MHD turbulence both at low (Zikanov & Thess 1998) and high (e. g. Biskamp & Müller 1999, Oughton et al 1994) magnetic Reynolds numbers, dynamo theory (Glatzmeier & Roberts 1995), plasma physics (Biskamp 1993) to applications including electromagnetic processing of materials (see the contributions by Garnier and Davidson in the present volume) and ferrofluids (Rosensweig 1985). In the present work we will limit our attention to incompressible electrically conducting fluids, thereby eliminating plasma physics and ferrofluids from our consideration.

The purpose of the present paper is to make the reader familiar with the basic physical and mathematical principles of CMHD and to illustrate the main sources of computational complexity. Moreover, the first part of our communication will provide a brief survey on recent work on CMHD. We will not attempt a comprehensive discussion of the mathematical and technical aspects of CMHD including grid generation, discretization, accuracy and implementation, neither does the limited available space permit us a discussion of all relevant publications in the field.

We rather provide two specific examples, one from fundamental CMHD (part II of the paper) and one from applied CMHD (part III of the paper) which, as we believe, give the reader an impression of the physical diversity and computational complexity of CMHD.

2. Principles of CMHD

2.1 Basic phenomena

The example of levitation melting discussed in the introduction (Fig. 1) permits one to understand the three main ingredients that are common to virtually all CMHD problems. As Fig. 2 shows, CMHD consists in general of an electrodynamic, a fluid dynamic part and (in most applications) a part describing the properties of the material or of a process that one is to understand and improve.

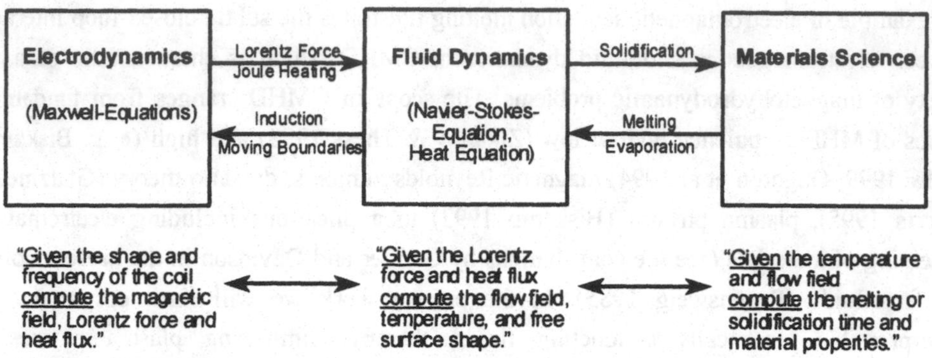

Figure 2: Basic phenomena modeled by computational MHD.

The complexity of CMHD comes from the fact that the complexity of each subdicipline is enhanced by a number of interactive processes sketched in Fig. 2. The electromagnetic field acts upon the flow field by two mechanisms, namely the Lorentz force

$$\mathbf{F} = \mathbf{J} \times \mathbf{B} \tag{2.1}$$

and by the heat flux

$$\dot{q} = \mathbf{J}^2 / \sigma \tag{2.2}$$

due to Joule heating. The back reaction of the fluid flow on the electromagnetic field occurs through the induction and is described by Ohm's law for moving conductors

$$\mathbf{J} = \sigma\,(\mathbf{E} + \mathbf{v} \times \mathbf{B}) \tag{2.3}$$

moreover, moving boundaries, solidification, melting, evaporation and wetting can modify the electromagnetic field.

2.2 Governing equations

The dynamical state of an electrically conducting fluid under the influence of an electromagnetic field is characterized in general by eight fields, namely three magnetic field components **B** (**r**, t), three velocity components **v** (**r**, t), pressure p (**r**, t), and temperature T (**r**, t). In frame of the Boussinesq approximation their evolution is governed by the magnetic field equation

$$\frac{\partial \mathbf{B}}{\partial t} + (\mathbf{v} \cdot \nabla)\mathbf{B} = (\mathbf{B} \cdot \nabla)\mathbf{v} + \frac{1}{\mu_0 \sigma} \nabla^2 \mathbf{B} \tag{2.4}$$

(derived from Maxwell's equations $\nabla \times \mathbf{E} = -\partial \mathbf{B} / \partial t$, $\mathbf{J} = \nabla \times \mathbf{B} / \mu_0$ and Ohm's law eq. 2.3), by the Navier-Stokes equation

$$\frac{\partial \mathbf{v}}{\partial t} + (\mathbf{v} \cdot \nabla)\,\mathbf{v} = -\frac{\nabla p}{\rho} + \upsilon \nabla^2 \mathbf{v} + \beta \mathbf{g}\,(T - T_0) + \frac{1}{\mu_0 \rho}\,(\nabla \times \mathbf{B}) \times \mathbf{B} \qquad (2.5)$$

and by the heat equation

$$\frac{\partial T}{\partial t} + (\mathbf{v} \cdot \nabla)\,T = \varkappa \nabla^2 T + \frac{1}{c_p \rho \sigma \mu_0^2}\,(\nabla \times \mathbf{B})^2 \qquad (2.6)$$

These equations are supplemented by the incompressibility condition $\nabla \cdot \mathbf{v} = 0$, by $\nabla \cdot \mathbf{B} = 0$ as well as by appropriate initial and boundary conditions. In view of the diversity of boundary conditions we shall only mention those which are most often encountered.

The magnetic field must be continuous across any discontinuity of the electrical conductivity such as e. g. the boundary of the melt in Fig. 1b. Moreover, the normal component of the electric current and the tangential component of the electric field have to be continuous as well. For the boundary conditions in the more complex case of ferromagnetic walls the reader is referred to Jackson (1962). Notice that the domains where equations (2.4) - (2.6) must be solved do not in general coincide. Whereas the Navier-Stokes equation has to be solved in the fluid only [Fig. 1], the magnetic field has to be computed in the entire space, while the temperature field is of interest only in the fluid. In most cases the fluid domain is bounded by the solid wall in which case the velocity field must obey the no-slip boundary condition $\mathbf{v} = 0$. For other types of boundary conditions such as free surfaces, fluid-fluid interfaces including the effect of non-uniform surface tension the reader is referred to fluid mechanics textbooks e. g. Landau & Lifshitz (1987).

The boundary conditions for temperature are either isothermal conditions $T = T_0$ or adiabatic conditions $\mathbf{n} \cdot \nabla T = 0$ or heat transfer boundary conditions $\lambda \mathbf{n} \cdot \nabla T = \alpha\,(T - T_\infty)$ with a phenomenological heat transfer coefficient α. For a discussion of the latter type of boundary conditions including methods for determining α the reader is referred to Incropera & DeWitt (1996).

Finally, it should be mentioned that the boundaries of the domains of solution are not fixed *a priori* but have to be determined as part of the solution procedure. The most important cases include free surfaces (like in Fig. 1) and solidification fronts (like in the course of the melting process from Fig. 1a to Fig. 1b). These problems require additional boundary conditions such as the kinematic boundary condition at the free surface (see Scardovelli & Zaleski 1999) or solidification models (Davis 1990).

It should be mentioned that the solution equations (2.4)-(2.6) is often based on alternative (but physically equivalent) mathematical formulations. Electromagnetic field computations often employ the representation of the magnetic and electric fields $\mathbf{B} = \nabla \times \mathbf{A}$, $\mathbf{E} = \nabla\phi_0 - \partial\mathbf{A}/\partial t$ in terms of magnetic (**A**) and electric (ϕ) potentials. Two-dimensional fluid flows, i. e. $\mathbf{v} = v_x(x, y, t)\mathbf{e}_x + v_y(x, y, t)\mathbf{e}_y$ are conveniently treated in terms of stream function ψ (x, y, t) which is defined as $v_x = \partial\psi/\partial y$, $v_y = -\partial\psi/\partial x$ and automatically satisfies the incompressibility constraints. Recently, a novel approach has been proposed (Meir & Schmidt 1994) based on equations for electric current density rather than magnetic field.

Although equations (2.4)-(2.6) represent the general system of basic CMHD equations, it is only under very rare circumstances that this full set of equation is really implemented in a numerical computation. The reason for this is a multitude of sources of complexity which will be discussed next.

2.3 Sources of complexity

The sources of complexity encountered when solving CMHD problems fall into two categories. First, we have the problems associated with each subdiscipline, namely computational electromagnetics, computational fluid dynamics and the models for solidification etc. Second, a number of additional difficulties arises from the interplay of the three aforementioned fields.

The complexity of electromagnetic field computations (eq. 2.4 or equivalent models, Binns et al 1992) arises primarily from the facts that

- the spatial resolution requirements become extremely high if the skin depth $\delta = (2\omega\mu_0\sigma)^{-1/2}$ [Fig. 1c] is much smaller than the dimension of the considend fluid,
- the space exterior to the fluid has to be meshed and included in the computation,
- the modeling of an imposed electric current in the coils leads to a non-standard boundary condition whose implementation is not straightforward,

- ferromagnetism and nonlinear dependence of material properties like electrical conductivity on T and **B** require to release the convenient assumption of sinusoidal time-dependence of **B** and perform fully transient simulations.

The complexity of the computation of flow fields (eq. 2.5 or equivalent models) is due to the observation that

- turbulent flows possess a tremendous disparity between the size of the smallest eddies ℓ and of the large eddies L approximately described by $\ell/L \approx Re^{-3/4}$ which makes a fully resolved simulation (a so called "direct numerical simulation") prohibitively expensive or even impossible for most practical purposes (see e. g. Moin & Mahesh 1998).

- flows with a free surface can develop regions with extremely high surface currature whose proper resolution becomes impossible in limit of vanishing surface tension. (Scardovelli & Zaleski 1999).

The sources of complexity associated with the dynamics of the solidification process are very diverse; for a comprehensive discussion the reader is referred to the literature (Davis 1990, Davis et al 1992).

In addition to the mentioned difficulties the interaction between the subtopics produces new challenges which are due to the fact that

- magnetohydrodynamic boundary layers, Hartmann layers, with thickness $\delta_h = B^{-1} (\sigma/\rho\nu)^{1/2}$ may arise in fluid flows under strong DC magnetic fields imposing severe spatial resolution requirements,

- time scales of the magnetic field and velocity field evolution characterised by the magnetic Prandtl number $Pm = \upsilon\mu_0\sigma$ are very different in many liquid metal flows ($Pm << 1$),

- the interactive dynamics between convection and solidification are still poorly understood.

Finally, it should be mentioned that complex geometry is a characteristic of many CMHD problems.

3. Survey of CMHD problems

To date, no single general-purpose code is available to solve the governing equations (2.4)-(2.6) in complex geometry. As a consequence, various approximations are adopted depending on the specific nature of the problem. In what follows we shall provide an overview of the most common CMHD problems ranging from engineering to geophysical applications.

3.1 Fluid flow in time-dependent magnetic fields

There are a wide and practically important class of CMHD problems in which the electromagnetic and fluid dynamic parts can be decoupled to the greatest possible extent (at least in a first approximation). These flows have fixed boundaries and are subjected to time-dependent magnetic fields. The three main types of magnetic fields are illustrated in Fig. 3.

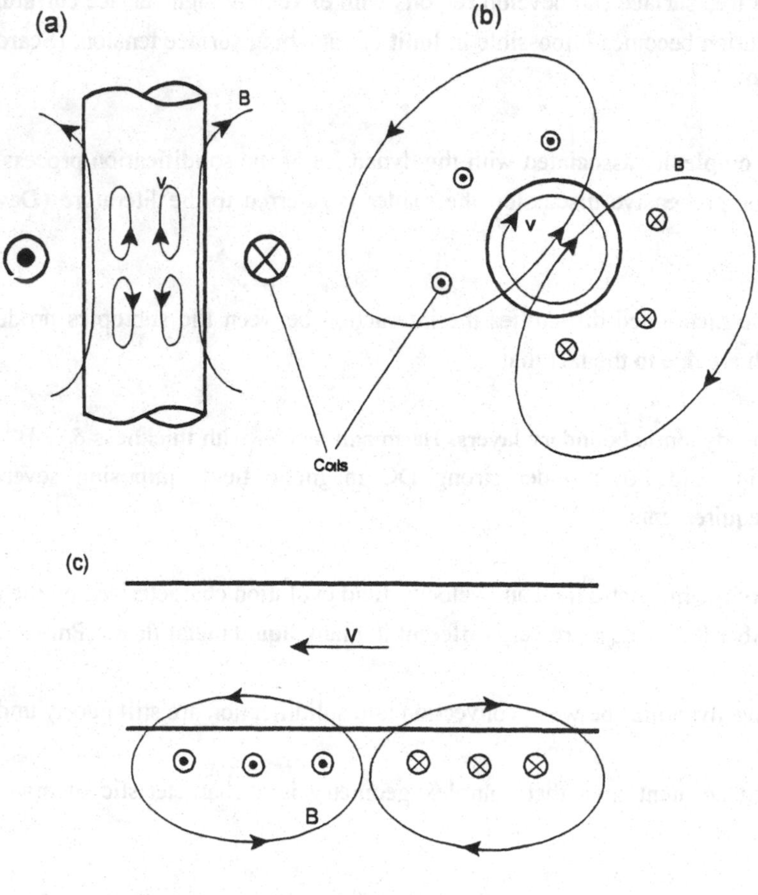

Figure 3: Fluid flow phenomena in time-dependent magnetic fields.
(a) heating and stirring by an oscillating magnetic field,
(b) stirring by a rotating magnetic field,
(c) stirring by a traveling magnetic field.

Fig 3a shows an oscillating field produced by a single coil acting on a cylindrical vessel filled with liquid metal. The field produces eddy currents and an inward-pointing Lorentz force whose strength is maximum in the mid-plane. As a result of the nonuniform Lorentz force a motion sets in which in the laminar case, consist of two toroidal vortices.

Figures 3b and 3c show a rotating and a travelling field respectively. Such fields are used for the electromagnetic stirring of steel and for induction pumps.

If the following assumptions are satisfied then the electromagnetic part of the problem can be decoupled form the fluid dynamic part:

(1) The magnetic Reynolds number $Rm = \mu_0 \sigma v L$ (where v is a characteristic velocity of fluid flow) obeys $Rm \ll 1$ so that the advective terms $(\mathbf{v} \cdot \nabla)\mathbf{B}$ and $(\mathbf{B} \cdot \nabla)\mathbf{v}$ in (2.4) can be neglected.
(2) The magnetic field has a sinusoidal time-dependence, i. e. $\mathbf{B}(\mathbf{r}, t) = \underline{\mathbf{B}}(\mathbf{r}) \cdot e^{j\omega t}$ which is the case in the absence of ferromagnetism.
(3) The magnetic field variation is much faster than the eddy turnover time $T = L/v$, i. e. $\omega^{-1} \ll T$. Then the Lorentz force and Joule heat flux can be replaced by their time-averages $\langle \mathbf{F} \rangle = \langle \mathbf{J} \times \mathbf{B} \rangle$ and $\langle \dot{q} \rangle = \sigma^{-1} \langle \mathbf{J}^2 \rangle$.

Under the foregoing assumptions the magnetic field can be computed as if the fluid were at rest because the resulting set of equations

$$j\omega \underline{\mathbf{B}} = \frac{1}{\mu_0 \sigma} \nabla^2 \underline{\mathbf{B}}, \quad \nabla \cdot \underline{\mathbf{B}} = 0 \qquad \text{in the fluid and coils} \qquad (3.1)$$

$$0 = \nabla \times \underline{\mathbf{B}}, \quad \nabla \cdot \underline{\mathbf{B}} = 0 \qquad \text{in the air or vacuum} \qquad (3.2)$$

does not contain the velocity field. This set of equations together with the conditions of continuity of the magnetic field, the normal component of electric current, the tangential component of electric field and $\underline{\mathbf{B}} = 0$ at infinity completely determines the magnetic field distribution.

Once $\underline{\mathbf{B}}(\mathbf{r})$ is known, $\langle F \rangle$ and $\langle \dot{q} \rangle$ are readily computed, and the fluid flow computation reduces to an "ordinary" computational fluid dynamics problem with prescribed force and heat flux, namely

$$\frac{\partial \mathbf{v}}{\partial t} + (\mathbf{v}\cdot\nabla)\mathbf{v} = \frac{-\nabla p}{\rho} + \upsilon\nabla^2\mathbf{v} + \langle \mathbf{F} \rangle \tag{3.3}$$

$$\frac{\partial T}{\partial t} + (\mathbf{v}\cdot\nabla)T = \varkappa\nabla^2 T + \frac{1}{c_p\rho}\langle \dot{q} \rangle \tag{3.4}$$

Part III of the present communication contains a detailed account of how the outlined procedure is applied to a specific engineering problem, namely electromagnetic control of melt flow in materials processing.

It should be stressed that even the reduced problem (3.3), (3.4) presents a formidable challenge to computational fluid dynamics since turbulence is still contained in this mathematical model. In computations for most engineering applications (characterized by high Reynolds number) recourse must therefore be had to turbulence models (Wilcox 1998) in which (3.3) is replaced by a set of model equations mimicking the time-averaged behavior of the turbulent velocity field.

Solutions of the system (3.1)-(3.4) for laminar time-independent flows or statistically-stationary flows with turbulence models have been computed quite early (Tacke & Schwerdtfeger 1979) in connection with applications of magnetic stirring of steel and continue to attract attention due to applications in other fields including growth of semiconductor crystals and vacuum are remelting (Vlasyuk & Sharamkin 1987, Muizhnieks & Yakovich 1988, Martin-Witkovski & Marty 1998). Computations of transient flows close to the critical Reynolds number for flow instability have recently been reported (Kaiser & Benz 1998, Friedrich et al 1999)

Computations of electromagnetically driven flows using large eddy simulation (LES - see Galperin & Orszag 1993) have recently begun to appear (Felten et al 1999) and will play an increasing role in future.

No direct numerical simulation (DNS - cf. Mahesh & Moin 1998) of a fully turbulent flow in one of the configurations sketched in fig. 2 is known to the authors. A problem in electromagnetic boundary layer control has been recently simulated (Crawford & Karniadakis 1997). It is expected that LES and DNS will greatly help in the next years to elucidate the complex nature of electromagnetically driven turbulent flows.

3.2 Fluid flows in a static magnetic field

A second class of generic CMHD problems pertains to the turbulent flow of an electrically conducting fluid under the influence of an externally applied static magnetic field. Fig. 4 shows the result of a classical experiment by Kolesnikov & Tsinober 1974 demonstrating that a homogeneous magnetic field brakes the flow of liquid metal and makes the turbulent flow strongly anisotropic. This problem is relevant at the electromagnetic braking of turbulent flows in steel production (cf. Davidson 1999) but represents also a fundamental problem in its own right.

The current state of direct numerical simulation is perhaps best characterized by an estimate showing that we are unlikely to see a direct numerical simulation of the seemingly simple problem of Fig. 4 before the year 2010 if the increase of computational power continues at its current pace.

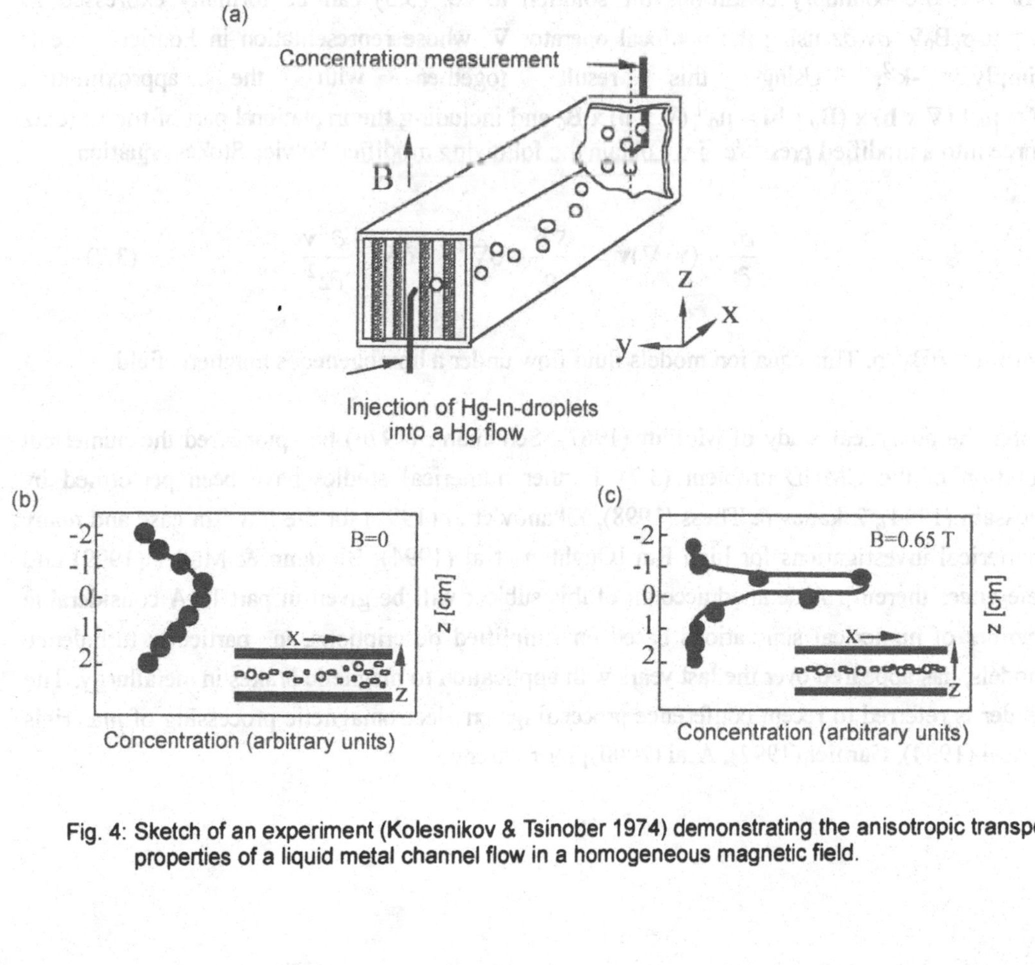

Fig. 4: Sketch of an experiment (Kolesnikov & Tsinober 1974) demonstrating the anisotropic transport properties of a liquid metal channel flow in a homogeneous magnetic field.

Instead of considering a wall-bounded flow it is often useful to study a simplified problem, namely the flow in a cubic box with periodic boundary conditions in all three directions. This simplification, referred to as homogeneous turbulence, permits the utilisation of highly efficient and accurate numerical methods, as will be described in detail in Part II of the present communication.

If the magnetic Reynolds number is small then equations (2.4) and (2.5) can be further simplified. Decomposing the magnetic field into an applied part $\mathbf{B}_0 = B_0\mathbf{e}_z$ and a perturbation **b**, i. e. $\mathbf{B} = \mathbf{B}_0 + \mathbf{b}$ with $|\mathbf{b}| << |\mathbf{B}_0|$ the magnetic field equation (2.4) simplifies to:

$$\nabla^2\mathbf{b} = \mu_0\sigma B_0 \frac{\partial \mathbf{v}}{\partial z}, \quad \nabla\cdot\mathbf{b} = 0 \quad \text{in the fluid} \qquad (3.5)$$

$$\nabla \times \mathbf{b} = 0, \qquad \nabla\cdot\mathbf{b} = 0 \quad \text{in insulating walls or vacuum} \qquad (3.6)$$

For periodic boundary conditions the solution to eq. (3.5) can be formally expressed as $\mathbf{b} = \mu_0\sigma_0 B_0 \nabla^{-2}\partial\mathbf{v}/\partial z$ using the nonlocal operator ∇^2 whose representation in Fourier-space is simply $-k^{-2}$. Using this result together with the approximation $\mathbf{F} = \mu_0^{-1}(\nabla\times\mathbf{b})\times(\mathbf{B}_0 + \underline{b}) \approx \mu_0^{-1}(\nabla\times\mathbf{b})\times\mathbf{B}_0$ and including the irrotational part of the Lorentz force into a modified pressure $\tilde{p}$ we obtain the following modified Navier-Stokes equation

$$\frac{\partial \mathbf{v}}{\partial t} + (\mathbf{v}\cdot\nabla)\mathbf{v} = -\frac{\nabla\tilde{p}}{\rho} + \upsilon\nabla^2\mathbf{v} - \alpha\nabla^{-2}\frac{\partial^2\mathbf{v}}{\partial z^2} \qquad (3.7)$$

with $\alpha = \sigma B_0^2/\rho$. This equation models fluid flow under a homogeneous magnetic field.

After the analytical study of Moffatt (1967), Schumann (1976) has pioneered the numerical solution of the CMHD problem (3.7). Further numerical studies have been performed by Hossain (1994), Zikanov & Thess (1998), Zikanov et al (1999) for the low Rm case and many numerical investigations for high Rm [Oughton et al (1994), Biskamp & Müller (1999) and references therein]. A detailed account of this subject will be given in part II. A considerable number of numerical simulations based on simplified descriptions, in particular turbulence models, has appeared over the last years with application to magnetic brakes in metallurgy. The reader is referred to recent conference proceedings on electromagnetic processing of materials [(Asai (1994), Garnier (1997), Asai (2000)] for references.

3.3 Problems involving strong coupling

The problems discussed in the previous two sections are characterized by the possibility to separate - to the greatest possible extent - the solution of the electrodynamic problem from the solution of the fluid-dynamics part. There are, however, a wide class of CMHD problems like the levitation problem sketched in Fig. 1 where electrodynamics and fluid dynamics are so intimately coupled that a separation is impossible. The present section will briefly discuss two representatives from this class of problems, namely nonlinear dynamo theory end the simulation of aluminium reduction cell instabilities.

Dynamo theory (for an introduction see the contribution of Moffatt to the present volume) is the science of generation and maintenance of magnetic fields by the movement of electrically conducting fluids. The numerical simulation of the evolution of the magnetic field of Earth, say, requires a numerical solution of the full set of equations (2.4)-(2.6), supplemented by an appropriate energy source - a task that has become feasible only recently (Fig. 5a). Such simulation (albeit with a simplified fluid flow model) has recently permitted, for the first time, to model the magnetic field reversals for the Earth (Glatzmaier & Roberts 1995).

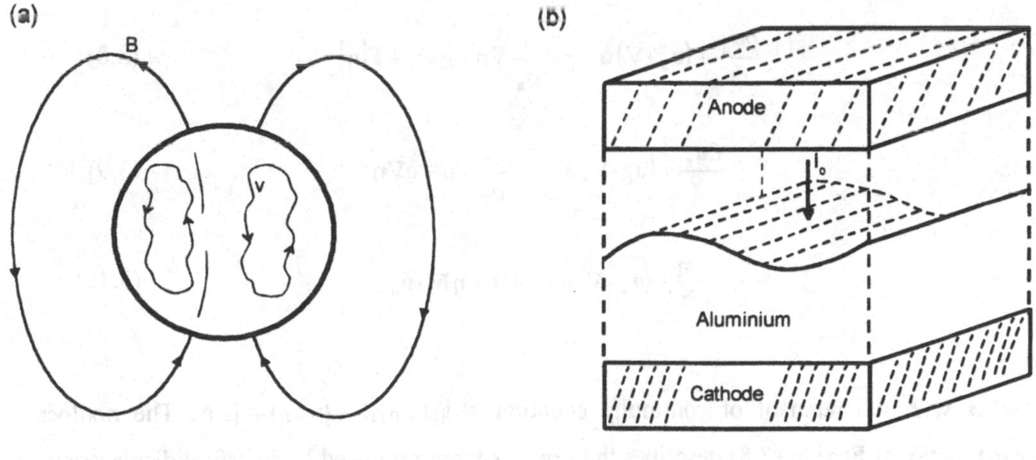

Figure 5: Examples of strongly coupled MHD-problems:
(a) Dynamo theory and flows at high magnetic Reynolds number,
(b) Interfacial instability in Aluminium reduction cells.

An example of a strongly coupled engineering CMHD problem is the instability in aluminium reduction cells. The physics of this intriguing problem have been discussed in detail by Davidson (1999). Suffice it here to say that the challenge is in a numerical prediction of the time-dependent interface (cf. Fig. 5b) between liquid aluminium and liquid cryolite whose unstable turbulent movement is set up by Lorentz forces produced by the interaction of a strong electric current passing through the cell with its own magnetic field. The complexity of the problem arises from the fact that the electric current density is determined by the shape of the interface. The latter is not known, however, until we have solved the dynamical evolution equations for both fluids. A second source complexity comes from the fact that the time-scales for the onset of unstable interface waves (hours) are much larger than those of turbulent movement (seconds). A "brute force" computation of the full 3D problem remains therefore out of reach for the foreseeable future. To make the problem numerically tractable Davidson & Lindsay (1998) and other authors have proposed a simplified yet self consistent model based on the shallow water approximation. This model assumes that the velocity fields in the aluminium u_a and cryolite u_c are predominantly two-dimensional and that the displacement η of the interface from its unperturbed shape z = 0 is much smaller than the depth H of the aluminium layer. Under these assumptions the evolution of the system is described by the equations:

$$\frac{\partial \mathbf{u}_a}{\partial t} + (\mathbf{u}_a \cdot \nabla)\mathbf{u}_a = -\frac{1}{\rho_a}\nabla p - g\nabla\eta + \mathbf{F}[\eta] \tag{3.8}$$

$$\frac{\partial \mathbf{u}_a}{\partial t} + (\mathbf{u}_c \cdot \nabla)\mathbf{u}_c = -\frac{1}{\rho_c}\nabla p - g\nabla\eta \tag{3.9}$$

$$\frac{\partial \eta}{\partial t} + (\mathbf{u}_a \cdot \nabla)\eta = -(\mathbf{H} + \eta)\nabla \cdot \mathbf{u}_a \tag{3.10}$$

together with the remnant of continuity equation $\nabla\cdot[(H+\eta)\mathbf{u}_a + (h-\eta)\mathbf{u}_c] = 0$. The nonlocal linear functional $\mathbf{F}[\eta]$ in (3.8) describes the Lorentz force produced by interface displacement. A numerical solution of eg. (3.8)-(3.10) has been obtained by Zikanov et al (2000).

4. Outlook

While considerable progress has been reached in the last years in the simulation of individual electromagnetic and fluid dynamic problems, CMHD is a young discipline with many interesting problems still to be done.

At the fundamental level we are likely to see soon direct numerical simulations (DNS) of homogenous MHD turbulence at resolutions 1024^3 which are very high by today's standards. DNS at lower resolution but including walls and full electromagnetics computation are likely to be possible for simple laboratory experiments like the one in Fig. 4 in the foreseeable future. Combination of DNS with computational techniques for free surface flows (Scardovelli & Zaleski 1999) will permit fascinating new insight into simple MHD free surface problems.

As the level of application the scope of numerical simulation is likely to broaden by using improved turbulence models (Widlund et al 1998) and by including effects like solidification, wetting and melting into consideration.

5. Acknowledgement

The authors are grateful to the Deutsche Forschungsgemeinschaft for financial support in frame of the Innovationskolleg "Magnetofluiddynamik".

References

Asai S. (Ed.) 1994: Proceedings of the First International Congress on Electromagnetic Processing of Materials, Nagoya 25-28 October 1994.

Asai, S. (Ed.), 2000: Proceedings of the 3rd International Symposium on Electromagnetic Processing of Materials, April 3-6, 2000, Nagoya, Japan; The Iron and Steel Institute of Japan.

Binns K.J.; P.J. Lawrenson; C.W. Trowbridge 1992: The Analytical and Numerical Solution of Electric and Magnetic Fields, John Wiley & Sons Ltd.

Biskamp, D., 1993: Nonlinear Magnetohydrodynamics, Cambridge University Press.

Biskamp, D.; W. C. Müller, 1999: Decay laws for three-dimensional MHD turbulence, Phys. Rev. Lett., vol. 83 (11), pp. 2195-2198.

Crawford C.H., G.E. Karniadakis 1997: Reynolds stress analysis of EMHD-controlled wall turbulence. Phys. Fluids vol. 9, 788.

Davidson P.A., R.I. Lindsay 1998: Stability of interfacial waves in aluminium reduction cells, J. Fluid Mech, vol. 362, pp. 273-296.

Davis, S. H., 1990: Hydrodynamic interactions in directional solidification J. Fluid Mech, vol. 212, pp. 241-262.

Davis, S. H.; H. E. Huppert; U. Müller; M. G. Worster (Eds.), 1992: Interactive Dynamics of Convection and Solidification, NATO ASI Series, vol. 219, Kluwer Academic Publishers, Dordrecht.

Egry, I.; Szekely J., 1991: The measurement of thermophysical properties in microgravity using electromagnetic levitation, Adv. Space Res., vol. 11(7), pp. 263-266.

Felten F., Y. Fautrelle, Y. Du Terrail, O. Metais 1999: Numerical modeling of electromagnetically driven flows using LES methods, in: Fluid Flow Phenomena in Metals Processing, Eds. N. El-Kaddah & D.G.C. Robertson, S.T. Johansen, V.R. Voller, The Minerals, Metals & Materials Society, pp. 571-579.

Friedrich J., Y.-S. Lee, B. Fischer, C. Kupfer, G. Müller 1999: Experimental and numerical study of Rayleigh-Benard convection affected by a rotating magnetic field, Phys. Fluids vol. 11. 853.

Galperin B., S.A. Orszag (Eds.) 1993: Large Eddy Simulation of Complex Engineering and Geophysical Flows, Cambridge University Press.

Garnier M. (Ed.) 1997: Proceedings of the 2nd International Congress on Electromagnetic Processing of Materials, Paris 27-29 May 1997.

Glatzmeier, G.; P. H. Roberts, 1995: Nature, vol. 377, pp. 203.

Hossain M. 1994: Reduction in the dimensionality of turbulence due to a strong rotation. Phys. Fluids. vol. 6, pp. 1077-1080.

Incropera, F. P.; D. P. DeWitt, 1996: Fundamentals of Heat and Mass Transfer, John Wiley & Sons. New York.

Jackson, J. D, 1962: Classical Electrodynamics, John Wiley & Sons, New York.

Kaiser, T.; K. W. Benz, 1998: Taylor vortex instabilities induced by a rotating magnetic field: A numerical approach. Phys. Fluids, vol. 10, 1104.

Kolesnikov Y.B., A.B. Tsinober 1974: Izv. Akad. Nauk SSSR - Mech Zhid I Gaza, vol 4, 146.

Landau, L. D.; E. M. Lifshitz, 1987: Fluid Mechanics, Course of Theoretical Physics, vol. 6, Pergamon Press.

Mahesh, K.; Moin, P., 1998: DNS: A tool in turbulence research, Annu. Rev. Fluid Mech. vol. 30, pp. 539-578.

Martin-Witkowski, L. M.; Ph. Marty, 1998: Effect of a rotating magnetic field of arbitrary frequency on a liquid metal column. Eur. I. Mech. B/Fluids, vol. 17, 239.

Meir, A. J.; P. G. Schmidt, 1994: A velocity - current formulation for stationary MHD flow, Appl. Math. Comp., vol. 65, pp. 95-109.

Moffatt H.K. 1967: On the suppression of turbulence by a uniform magnetic field, J. Fluid Mech, vol. 28, pp. 571-592.

Muizhnieks A.R., A.T. Yakovich 1988: Numerical study of closed axisymmetric MHD rotation in an axial magnetic field with mechanical interaction of azimuthal and meridional flows, Magnetohydrodynamics, vol. 24, pp. 50-55.

Oughton, S.; E. R. Priest; W. H. Matthaeus, 1994: The influence of a mean magnetic field on three-dimensional MHD turbulence, J. Fluid Mech., vol. 280, pp. 95-117.

Rosensweig, R. E., 1985. Ferrohydrodynamics, Cambridge University Press.

Scardovelli, R.; S. Zaleski, 1999: Direct numerical simulation of free-surface and interfacial flow, Annu. Rev. Fluid Mech., vol. 31, pp. 567-603.

Schumann U. 1976: Numerical simulation of the transition from three- to two-dimensional turbulence under a uniform magnetic field, J. Fluid Mech, vol. 74, pp. 31-58.

Tacke K.H.; K. Schwerdtfeger 1979: Stirring velocities in continuously cast round billets as induced by rotating electromagnetic fields, Stahl und Eisen, vol. 99, pp. 7-12.

Vlasyuk V.K.; V.I. Sharamkin 1987: Effect of vertical magnetic field on heat and mass transfer in a parabolic liquid metal bath, Magnetohydrodynamics, vol. 23, pp. 211-216.

Widlund O., S. Zahrai, F. Bark 1998: Development of a Reynolds stress closure for modeling of homogeneous MHD turbulence, Phys. Fluids, vol. 10, 1987.

Wilcox D.C 1998: Turbulence modeling for CFD, DCW Industries.

Zikanov, O.; A. Thess, 1999: Direct numerical simulation of forced MHD turbulence at low magnetic Reynolds number, J. Fluid Mech., vol. 358, pp. 299-333.

Zikanov O., D. Ziegler, P.A. Davidson, A. Thess 2000: Metall Trans. (in press).

Chapter V

E. Zienicke, H. Politano, A. Pouquet

COMPUTATIONAL MHD
PART II -
APPLICATION TO A NONLINEAR DYNAMO MODEL

Computational MHD
Part II – Application to a Nonlinear Dynamo Model

Egbert Zienicke[1], Hélène Politano[2], Annick Pouquet[2]

[1] Universität Ilmenau, Fakultät Maschinenbau, Germany
[2] Observatoire de la Côte d'Azur, Nice, France

Abstract. Lagrangian chaos of the underlying flow is the driving force for the fast dynamo based on the strectch-twist-fold mechanism on small scales. In this contribution the hypothesis that the magnetic field by the action of the Lorentz force supresses Lagrangian chaos is checked by direct numerical simulations of the MHD equations. As a measure of the level of chaos the Lyapunov exponent of a set of 128×128 trajectories of fluid particles is computed in the growth phase and in the saturated phase of the dynamo when the magnetic field has reached its final strength. The numerical code, based on a pseudospectral algorithm, is developped for parallel computation on a multiprocessor system. Magnetic Reynolds numbers up to 240 and scale separations between the wavelength of the hydrodynamical forcing and the scale of the computational domain up to four are reached. For the runs were the kinetic Reyold number is high enough that the hydrodynamical bifurcation sequence to a more chaotic flow already has taken place, the mean value of the Lyapunov exponent is noticeable diminished in the saturated phase compared to the growth phase of the dynamo.

1 Introduction

As observed in Geophysics, Planetology and Astronomy different cosmical objects as planets, stars or the galaxy as a whole own magnetic fields. What all this bodies have in common is the presence of an electrically conducting fluid. This may be a fluid in a metallic state (for example liquid iron in a spherical shell around the Earth core or liquid, metallic hydrogen inside of Jupiter) or a plasma (as in stellar athmospheres or the intergalactic medium). Comparing the decay times for this fields with the times of their existence one is led to the conclusion that a mechanism is needed which is able to explain how the magnetic field of a body arises and how it is maintained. This mechanism is delivered by the so called *dynamo effect*, which means that for a given flow an arbitrary small seed magnetic field is exponentially amplified by the flow through induction effects (see also the introduction to dynamo theory by Moffat in the present volume).

A quantity of electrically conducting fluid with a given flow $\mathbf{u}(\mathbf{x}, t)$ inside of a bounded region is a dynamo, if an arbitrary small magnetic field perturbation $\delta\mathbf{B}$ is growing exponentially, i. e. if the state $\mathbf{B} \equiv 0$ is unstable against small perturbations. This is the statement of the *linear* or *kinematic dynamo problem*, where only the origination of a magnetic field is considered, but not how the magnetic field saturates to its final state. The term 'linear' stems from the fact that the induction equation, giving the law for the time evolution of the magnetic field, is linear in $\mathbf{B}$ for a given flow:

$$\frac{\partial \mathbf{B}}{\partial t} = \nabla \times (\mathbf{u} \times \mathbf{B}) + \eta \nabla^2 \mathbf{B}. \tag{1}$$

Here $\eta = 1/\mu_0\sigma$ is the magnetic diffusivity. The linear dynamo problem has been investigated intensively in the past, theoretically as well as numerically on model systems. Two main mechanisms, which are valid for high Reynolds number flows and have both already an extended theory behind, are often quoted: (1) the *turbulent dynamo* based on *helical turbulent fluctuations* of the underlying three-dimensional flow (for reviews see Moffatt (1978), Krause and Rädler (1980) and Roberts and Sowards (1992)) and (2) the *fast dynamo* based on the stretch-twist-fold mechanism on small scales provided by *chaotic motion* of the underlying flow (a review especially for the fast dynamo is given by the book of Childress and Gilbert (1995).

The linear phase of a dynamo lasts as long as the magnetic field is so small that the Lorentz force $\mathbf{j} \times \mathbf{B}$ is not able to change the velocity field. But, as the magnetic field is growing exponentially in the linear phase, one has only to wait long enough that a realizable backreaction on the velocity field will take place. This is the beginning of the *nonlinear phase* were a saturation of the magnetic field will take place in the end by means of the action of the Lorentz force on the velocity field and by means of Ohmic dissipation. To describe the saturation one has to solve the full magnetohydrodynamical equations, this means the Navier-Stokes equation with Lorentz force *and* the induction equation:

$$\frac{\partial \mathbf{u}}{\partial t} + (\mathbf{u}\cdot\nabla)\mathbf{u} = -\frac{1}{\rho}\nabla p + \nu\Delta\mathbf{u} + \frac{1}{\rho}\mathbf{j}\times\mathbf{B} + \mathbf{f}, \tag{2}$$

$$\frac{\partial \mathbf{B}}{\partial t} + (\mathbf{u}\cdot\nabla)\mathbf{B} = (\mathbf{B}\cdot\nabla)\mathbf{u} + \eta\Delta\mathbf{B}. \tag{3}$$

Because of the Maxwell equation $\mathbf{j} = \nabla \times \mathbf{B}/\mu$ the Lorentz force also is a nonlinearity in the Navier-Stokes equation, additionally to the inertia term $(\mathbf{u}\cdot\nabla)\mathbf{u}$. The fact that the nonlinearities in the Navier-Stokes equation as well as in the induction equation are not negligible is the reason that direct numerical simulations are indispensable for the investigation of the saturation process to test hypotheses about saturation mechanisms which are received by other means.

There are several attempts to find models for a better understanding of the back reaction of the magnetic field on the velocity field and to derive mechanisms for the process of saturation. Using closures of turbulence Pouquet et al. (1976) and Bhattacharjee and Yuan (1995) found, that saturation can occur in the presence of a large scale magnetic field through helical Alfvén waves. The large scale field originates from an inverse cascade of magnetic helicity, which is due to the invariance of the total helicity $H^M = <\mathbf{B}\cdot\nabla\times\mathbf{B}>$. Models (Pouquet et al. (1976), Frisch et al. (1975)) and computations in the incompressible (Pouquet and Patterson (1978), Meneguzzi et al. (1981)) and compressible cases — both subsonic (Horiuchi and Sato (1988)) and supersonic (Balzara and Pouquet) — indicate linear growth of magnetic helicity, including in the nonlinear dynamic regime.

An important role in saturation is played by a large scale magnetic field. A strong homogenous field (largest possible scale) is known to suppress turbulent fluctuations parallel to the magnetic field lines. This is of importance, both for the turbulent as well as the fast dynamo. Antidynamo theorems suggest, that an inherent three-dimensional motion of the flow is necessary for dynamo action, whereas a strong magnetic field has the tendency to force the flow into a two-dimensional structure. Helical fluctuations (necessary for the turbulent dynamo) as well as stretch-twist-fold operations at small scales (as necessary for the fast dynamo) are only possible in three dimensions. Nevertheless, the magnetic field developping in a dynamo process normally does not have the simple structure of a homogenous field. Often a strong filamentary and intermittent structure of the growing magnetic field is observed in astrophysical flows (for example in the sun athmosphere) as well as in numerical simulations (see Galanti et al. (1992), Brandenburg et al. (1996), Galloway and Frisch (1986)). If the magnetic field is localized in strong flux tubes the suppression of small scales of flow also would be localized therein, whereas in the rest of the volume the flow would be only weakly influenced and thus still be able to produce magnetic field on small scales. The overall strength of the suppression of magnetic field production in this case would be strongly influenced by the spatial structure of the magnetic field (see Blackman (1996)).

Instead of going into the details of spatial structure in this work we focus on prooving the effect as it should be stated for the fast dynamo: *wether a growing magnetic field is able to suppress Lagrangian chaos of the flow on small scales.* This interesting question has been raised by Cattaneo et al. (1996) and was investigated numerically on a dynamo model with strong confinements on the development of the nonlinear terms of the flow: (i) The advection term responsible for turbulence in an incompressible fluid is totally neglected, allowing for a simple decomposition of the linear and nonlinear phases of the dynamo, (ii) The Lorentz force is averaged in z-direction to let the flow retain a two-dimensional (but timedependent) structure. With these assumptions, a clear diminution of chaos — as diagnosed by a two-dimensional map of finite-time Lyapunov exponents — is received. As the nonlinearity, which is supposed to suppress Lagrangian chaos is present in these calculations (although in reduced form), we find this result encouraging. On the other hand, the neglection of the advection term cannot be without influence on the development of the velocity field, especially on the *development of small scales promoting turbulence and also Lagrangian chaos.* We therefore investigate the same problem on a different dynamo model, solving the full MHD-equations in three dimensions and dropping all simplifying assumptions.

In the next section we introduce the dynamo model which will be investigated. Section 3 gives some detail about the numerical implementation of the three-dimensional, parallel MHD-solver and how Lyapunov exponents are calculated in the code to measure the level of Lagrangian chaos. Section 4 is devoted to the presentation of the numerical results.

2 The Model

To avoid all complications arising from wall bounded flow the fluid is situated inside a box of side length L and periodic boundary conditions are applied in all three space directions. This also allows for the application of pseudo spectral methods having the advantage of high precision for the direct numerical simulation (see Gottlieb and Orzag (1977) and Canuto et al. (1988)). The MHD-equations are computed in Alphénic units, this means the magnetic field has also the dimension of a velocity by the transformation $\mathbf{b} = \mathbf{B}/\sqrt{\rho\mu}$. The equations then read

$$\frac{\partial \mathbf{u}}{\partial t} + (\mathbf{u}\cdot\nabla)\mathbf{u} = -\nabla P + \nu\Delta\mathbf{u} + (\mathbf{b}\cdot\nabla)\mathbf{b} + \mathbf{f}_{ABC}, \tag{4}$$

$$\frac{\partial \mathbf{b}}{\partial t} + (\mathbf{u}\cdot\nabla)\mathbf{b} = (\mathbf{b}\cdot\nabla)\mathbf{u} + \eta\Delta\mathbf{b}, \tag{5}$$

where we have included the magnetic pressure into the total pressure $P = p/\rho + \mathbf{b}^2/2$. Additionally, the magnetic field and the velocity field fulfill $\nabla\cdot\mathbf{b} = 0$ and $\nabla\cdot\mathbf{u} = 0$, the latter because we assume the fluid to be incompressible. Although our numerical investigation mainly is aimed at the fast dynamo we nevertheless have chosen a forcing which additionly to chaos of the underlying flow gives rise also to helical fluctuations with the possibility of scale separation. The so called ABC-forcing is given by

$$\mathbf{f}_{ABC} = \nu k_0^2 \mathbf{u}_{ABC} = \nu k_0^2 \begin{pmatrix} A \sin k_0 z + C \cos k_0 y \\ B \sin k_0 x + A \cos k_0 z \\ C \sin k_0 y + B \cos k_0 x \end{pmatrix}. \tag{6}$$

The ABC-flow $\mathbf{u}_{ABC}$ consists of three orthogonal Beltrami waves with amplitudes A, B and C. The wavenumber k_0 allows to introduce a scale separation between the forcing and the largest lengthscale, which is the length of the box. ABC-flows are exact solutions of the Euler equation and also of the Navier-Stokes equation, if the above forcing $\mathbf{f}_{ABC}$ is applied. Because of the Beltrami property $\nabla \times \mathbf{u}_{ABC} \propto \mathbf{u}_{ABC}$ they have strong helicity. This is the reason why they early were studied in the context of the α-effect (see Childress (1970)).

If all three amplitudes are non-zero the flow is non-integrable (in the sense of the theory of dynamical systems) and shows a mixture of regular islands and chaotic regions in a Poincaré-section (see Arnold (1965), Henon (1966) and Dombre et al. (1986)). Because of the existence of Lagrangian chaos in the flow the ABC-dynamo is a candidate for fast dynamo action. This question was investigated in some numerical studies in the framework of the linear dynamo theory: (Arnold and Korkina (1983), Galloway and Frisch (1984), Galloway and Frisch (1986) and Lau and Finn (1993)). The results are positive, but the highest magnetic Reynolds number reached are about 10^3, so that the answer is not entirely decisive.

ABC-flows without magnetic field were investigated by Podvigina and Pouquet (1994). For the case $A = B = C = 1$ the ABC-flow is a stable solution of the Navier-Stokes equation up to a Reynolds number about 13. At a critical Reynolds number of 13.04 a bifurcation to a timedependent solution takes place. Further bifurcations in a regime still at low Reynolds number are observed. The flow which evolves under ABC-forcing becomes more and more turbulent for higher Reynolds number.

The nonlinear dynamo with ABC-forcing has already been investigated under different questions: In Galanti et al. (1992) among other topics scaling effects by varying the parameter k_0 between 1, 2 and 4 are studied for the nonlinear dynamo. It is found that the level of saturation increases significantly by increasing k_0. Feudel et al. (1996a) and Feudel et al. (1996b) investigate the bifurcation structure of the nonlinear dynamo for low kinetic and magnetic Reynolds numbers (up to $R^V = R^M = 20$) and classify the symmetries of the solutions they find.

3 Numerical Implementation

The code for the direct numerical simulation is developped as a parallel code running on a multiprocessor system (Cray-T3E). As already mentioned in the last section it is of spectral type using expansions into exponential functions for the space dependence of $\mathbf{u}(\mathbf{x}, t)$ and $\mathbf{b}(\mathbf{x}, t)$:

$$\mathbf{u} = \sum_{\mathbf{k}} \mathbf{u}_{\mathbf{k}} e^{i\mathbf{k}\cdot\mathbf{x}}, \qquad \mathbf{b} = \sum_{\mathbf{k}} \mathbf{b}_{\mathbf{k}} e^{i\mathbf{k}\cdot\mathbf{x}}, \tag{7}$$

where $\mathbf{u}_{\mathbf{k}}$ and $\mathbf{b}_{\mathbf{k}}$ are the time-dependent expansion coefficients giving the amplitudes for the modes $\mathbf{k} = (2\pi/L)\,(n_1, n_2, n_3)^T$ in Fourier space (n_1, n_2 and n_3 denote integer numbers). The partial differential equations (4) and (5) are in this way represented by an infinite number of ordinary differential equations for the Fourier coefficients:

$$\dot{\mathbf{u}}_{\mathbf{k}} = \mathbf{w}_{\mathbf{k}} - i\mathbf{k}P_{\mathbf{k}} - \nu k^2 \mathbf{u}_{\mathbf{k}} + \mathbf{f}_{\mathbf{k}}^{ABC}, \tag{8}$$

$$\dot{\mathbf{b}}_{\mathbf{k}} = \mathbf{c}_{\mathbf{k}} - \eta k^2 \mathbf{b}_{\mathbf{k}}. \tag{9}$$

$\mathbf{w}_{\mathbf{k}}$ and $\mathbf{c}_{\mathbf{k}}$ denote the Fourier coefficients of the nonlinearities of the Navier-Stokes equation and of the induction equation, respectively:

$$w_j = b_i \frac{\partial b_j}{\partial x_i} - u_i \frac{\partial u_j}{\partial x_i} = \frac{\partial}{\partial x_i}(b_i b_j - u_i u_j), \tag{10}$$

$$c_j = b_i \frac{\partial u_j}{\partial x_i} - u_i \frac{\partial b_j}{\partial x_i} = \frac{\partial}{\partial x_i}(b_i u_j - u_i b_j). \tag{11}$$

The second form of the nonlinearity, often called conservative form, is received using $\nabla \cdot \mathbf{u} = \nabla \cdot \mathbf{b} = 0$.

The pressure in the Navier-Stokes equation (8) can be eliminated in Fourier space. Taking the divergence of the Navier-Stokes equation one gets for the pressure $P_{\mathbf{k}} = -i\mathbf{k} \cdot \mathbf{w}_{\mathbf{k}}/k^2$, what can be inserted back again into (8) with the result:

$$\dot{v}_{\mathbf{k}}^{(j)} = \left(\delta_{ij} - \frac{k_i k_j}{k^2}\right) w_{\mathbf{k}}^{(i)} - \nu k^2 v_{\mathbf{k}}^{(j)} + f_{\mathbf{k}}^{ABC,(j)} \tag{12}$$

Equations (12) and (9) are integrated using an Adams-Bashforth/Crank-Nicolson timestep, i.e. an explicit timestep of second order for the nonlinear terms and the forcing and an implicit timestep of second order for the diffusion terms.

The computation of the nonlinearities in each timestep would need $o(N^6)$ operations if they would be computed directly in Fourier space. This operation count can be reduced significantly by using a pseudo spectral method (see 18 (18) and 7 (7)). The nonlinearity is first transformed into physical space by a fast Fourier tranformation needing $o(N^3 \log(3N))$ operations. There it can be computed pointwise by $o(N^3)$ operations. Finally the result is transformed back by a second FFT into the Fourier space again. The leading order in the expression for the total number of operations for each timestep then is determined by the Fast Fourier transformation.

The crucial step in parallelizing a code is to find the best distribution of the data onto the processors allocated for the computation in order to avoid time consuming communication between processors. For the MHD-solver this concerns mainly the distribution of the 3d-arrays for the velocity and the magnetic field as well in Fourier space as in physical space. The psuedo spectral algorithm is well suited for parallelization. The timestep itself and the computation of the nonlinearities can be done by each processor on its own data. Communication between different processors only is necessary in the Fast Fourier transformations. But for this task optimized parallel library routines are available on the Cray. To use the parallelized FFT's in a Fortran code the second and third dimensions of the 3d-arrays are distributed on a two-dimensional array of processors. In our implementation the total number of processors has to be a power of 2 as is also the case for the spatial resolution N in the three space dimensions. Were communication was necessary, especially in input and output routines, the *shmem*-subroutines of the Cray-T3E were used. With 64 processors the code needs the following computation times for one timestep: 0.11s for a resolution of 64^3 gridpoints, 0.68s for resolution 128^3 and 6.5s for resolution 256^3.

As a measure of stretching we compute the largest Lyapunov exponent of the flow in the growth phase and in the saturated phase of the dynamo. The Lyapunov exponent measures the exponential growth in time of the distance between two initially nearby fluid particles. Denoting by $\mathbf{x}_0$ and $\mathbf{x}_0 + \delta\mathbf{x}$ the initial positions of the fluid particles at time t_0 the Lyapunov exponent is defined as

$$\lambda(\mathbf{x}_0) = \lim_{t\to\infty} \frac{1}{t} \log\left(\frac{|\mathbf{x}(t;\mathbf{x}_0,t_0) - \mathbf{x}(t;\mathbf{x}_0+\delta\mathbf{x},t_0)|}{|\delta\mathbf{x}|}\right) \tag{13}$$

If the flow is exponential stretching the Lyapunov exponent is greater than zero, else it is equal to zero. To compute the Lyapunov exponent for a starting point $\mathbf{x}_0$ one has to integrate the trajectories $\mathbf{x}(t;\mathbf{x}_0,t_0)$ and $\mathbf{x}(t;\mathbf{x}_0+\delta\mathbf{x},t_0)$ of the corresponding fluid particles by integrating the three-dimensional ordinary differential equation

$$\dot{\mathbf{x}}(t;\mathbf{x}_0,t_0) = \mathbf{u}(\mathbf{x}(t;\mathbf{x}_0,t_0),t). \tag{14}$$

To get an overwiev of the Lyapunov exponents in the whole phase space we start two neighbouring trajectories at 128 ×128 equally spaced starting points $\mathbf{x}_0^{(i,j)}$ in an intersection plane $x = const$ of the computational domain at a chosen time. All these trajectories are followed for a finite time and the finite time Lyapunov exponent is computed numerically (for more details see Guckenheimer and Holmes (1983) and Ott (1993)). To be mathematically correct one would have to integrate the trajectories an infinite time, however at least in the growth phase of the dynamo we are limited from above by the time when the magnetic field begins to saturate and the kinematic dynamo phase ends. We compare with Lyapunov exponents computed for a time interval of the same length in the saturated phase.

The computation of the Lyapunov exponents is included in the MHD-code integrating velocity field and magnetic field in time. Parallel to the timestep of the MHD-solver all 128 ×128 trajectories are advanced in a timestep of the diffential equation (14) using the actual velocity field to calculate the right hand side. All 128×128 trajectories are equally distributed over the available processors so that each processor has to compute the right hand side for his trajectories. The FFT delivers the values of the velocity in physical space only on a cartesian grid of points, while the trajectories reach arbitrary points in the computational domain. Therefore the right hand side is computed directly from the Fourier expansion (7) using the known Fourier coefficients. This needs $O(N^3)$ operations for trajectories in each timestep. This is a lot, especially because of necessary communication between processors (the Fourier coefficients are also on different processors). Therefore, for the computation of the Lyapunov exponents a harder truncation at a wavenumber k_M is used as for the MHD-solver. For suffiently low k_M the corresponding Fourier coefficients are stored only on a few processors (ideally four) and can be broadcasted to all other processors for the computation of the right hand side. Using all modes up to the shell $k_M = 15$ we got the following computation time for one timestep with computation of Lyapunov exponent for 128 ×128 trajectories: 21.5s with 16 processors at resolution 64^3 ($\approx$ factor 53 compared to MHD-solver running alone) and 6.5s with 64 processors at resolution 128^3 ($\approx$ factor 9 compared to MHD-solver running alone).

4 Numerical Results

Before we begin with the presentation of the results we first define the lengthscale and the timescale, in which we express our data. The length of the computational domain is chosen as $L = 2\pi$. Therefore k_0 is an integer number and equal to the scale separation $L/l_0 = L/(2\pi/k_0) = k_0$, where l_0 denotes the wavelength of the forcing. We define the kinetic and the magnetic Reynolds numbers corresponding to the length $L^* = l_0/2\pi$ (the length scale of the forcing divided by 2π) and to the velocity $U^* = \sqrt{(A^2+B^2+C^2)/3}$ (by this choice one has $U^* = 1$ for $A = B = C = 1$):

$$R^V = \frac{\sqrt{(A^2+B^2+C^2)/3}}{k_0\nu}, \quad R^M = \frac{\sqrt{(A^2+B^2+C^2)/3}}{k_0\eta}. \tag{15}$$

This is the same definition as was used in Galanti et al. (1992). All times in the following are measured in units of the turnover time

$$\tau_{NL} = \frac{L^*}{U^*} = \frac{l_0/2\pi}{U} = \frac{1}{k_0\sqrt{(A^2+B^2+C^2)/3}}. \tag{16}$$

In this time scaling the value of the Lyapunov exponent for pure ABC-flow will scale by a factor of k_0 for different scale separations because of the factor $1/t$ in its definition (13). As a first check of the Lyapunov part of the code and to have a comparison for the following calculations in the growth phase and in the saturation phase of the nonlinear dynamo we computed the Lyapunov exponents for the pure ABC_{k_0}-flow for $k_0 = 1, 2$ and 4. We started with 128×128 equally distributed initial conditions in the plane $x = \pi/2$. In figure 3 and 5 one can see the result for the ABC_1-flow after 80 timeunits and for the ABC_2-flow after 160 timeunits as grey scale image: bright regions correspond to small Lyapunov exponents and dark regions to high Lyapunov exponents. There are small bands of chaos visible beloining to a chain of hyperbolic fixpoints. Figure 3 can be compared with figure 10 of Dombre et al. (1986). For $A = B = C$ the chaos is rather weak taking the mean value λ (ABC_{k_0}), because the larger part of phase space corresponds to integrable respective nearly integrable flow (KAM-tori). Convergence of the mean Lyapunov exponent is reached for a time interval greater thean 800 (changement less than 4%). For $\Delta t = 160$ the mean values for the LE are 0.0243 for $k_0 = 1$, 0.0536 for $k_0 = 2$ and 0.1251 for $k_0 = 4$, which confirms roughly our statement above: the ratio should be 1:2:4 and actually is 1:2.21:5.15. With the exception of run 1 the flow is different from the ABC-flow in the growth phase and the saturated phase of the dynamo, and the convergence is better than for the pure ABC-flow.

We calculated Luapunov exponents for four different runs with scale separations $k_0 = 1, 2$ and 4 and different kinetic and magnetic Reynolds numbers. For the amplitudes A, B, and C of the forcing we restricted ourselves to the case of equal amplitudes $A = B = C = 1$, which is the best investigated up to now (see Galloway and Frisch (1986), Galanti et al. (1992)). The runs are listed in table 1.

The procedure how the magnetic seed is introduced is different for run 1 compared to the other runs. The reason is that the kinetic Reynolds number for run 1 is still in a regime where the ABC_1-flow is a stable solution of the pure hydrodynamical equation. Therefore run 1 is started with the initial condition $\mathbf{u} = \mathbf{u}^1_{111} + \delta\mathbf{u}$ for the velocity and $\mathbf{b} = \delta\mathbf{b}$ for the magnetic field, where $\mathbf{u}^1_{111}$ is the ABC-flow for $k_0 = 1$ and $\delta\mathbf{u}$ and $\delta\mathbf{b}$ are small perturbations with an energy about 10^{-9} in the first six shells of k-vectors. The instability of the ABC-flow in this run arises only because of the presence of the magnetic field perturbation.

For the three other runs we demonstrate the procedure of setting the magnetic seed on run 2 (see figure 1). The ABC-flow now is hydrodynamically unstable. To separate between the purely

run	k_0	R^V	R^M	N	χ	Δt	$\bar{\lambda}_g$	$\bar{\lambda}_s$	$\bar{\lambda}_s/\bar{\lambda}_g$
1	1	12	12	16	0.03	320	0.019	0.053	2.74
2	1	60	240	64	0.12	80	0.115	0.073	0.64
3	2	60	240	128	1.0	160	0.189	0.090	0.48
4	4	12	12	64	1.4	160	0.437	0.256	0.59

Table 1. List of runs. The first four columns give the scale separation, the Reynolds numbers and the resolution for the runs. $\chi = E^M/E^V$ gives the ratio of magnetic to kinetic energy in the saturated phase of the nonlinear dynamo. In the last four columns are listed the following data: the time intervall for the computation of the Lyapunov exponent, the mean value of the Lyapunov exponent for 128 ×128 trajectories in the growth phase and the saturated phase, and the ratio of mean value of Lypunov exponents between the saturated phase and the growth phase.

hydrodynamic instability and the dynamo effect we first let the ABC-flow destabilize in a pure hydrodynamic run with initial condition $\mathbf{u} = \mathbf{u}^1_{111} + \delta\mathbf{u}$ for the velocity and zero magnetic field $\mathbf{b} \equiv 0$. The perturbation grows exponentially until the ABC-flow (shell 1 in figure 1) breaks down (before timeunit 50 in figure 1) to a time dependent state with all modes excited. This ABC-forced state is the flow field were now the magnetic seed $\delta\mathbf{b}$ is introduced (timeunit 150 in figure 1) and the growth phase of the dynamo begins. When the magnetic field strength is high enough the nonlinear phase begins and one can see, that the velocity field is influenced by the magnetic field (timeunit 350 to 450 in figure 1). Then the nonlinear dynamo saturates, i.e. the magnetic field and the velocity field reach their final state.

We now turn to a more detailed description of the results for the individual runs. As already mentioned in the growth phase for run 1 the flow is very near to the ABC-flow. This expresses itself in an only slighthy higher value of the mean Lyapunov exponent than for pure ABC-flow. The finite time Lyapunov exponent is computed for the time intervals [0, 320] in the growth phase and [2920, 3240] in the saturated phase. The level of saturation measured by $\chi = E^M/E^V$ is about 3% for this run. In this run the dynamo bifurcation *enhances* the level of chaos from the growth phase to the saturation phase. In figure 2 one can see, that besides a peak of very low Lyapunov exponent there is a long tail of higher Lyapunov exponent values in the saturated phase. This result can be explained as follows: The ABC-flow is a stationary flow and the flow for the saturated dynamo is periodic (i.e. the bifurcation to the magnetic state is a Hopf bifurcation). A time dependent flow of course generates more chaotic trajectories than a stationary flow. Run 1 thus can not be considered as generic concerning the question of suppression of chaos by a growing magnetic field out of the following reasons: (i) Reynolds number is low, (ii) the flow is stationary and not in the turbulent regime, i.e. the peculiarities of the dynamics in the transitional regime influence the result, (iii) the magnetic field still is weak, so the effect is rather due to pure hydrodynamical effects (that appear because the ABC-flow is destabilized by the magnetic field) than the action of the Lorentz force.

Keeping the scale separation but increasing the kinetic and magnetic Reynolds numbers to $R^V = 60$ and $R^M = 240$ run 2 was performed. The distribution of Lyapunov exponents is now represented in figure 2 and figure 3 in two different ways. In figure 2 the histogram of Lyapunov exponents now shows a Gaussian distribution with a mean of 0.115 in the growth phase. That the flow in the growth phase has a much higher level of Lagrangian chaos than the pure ABC-

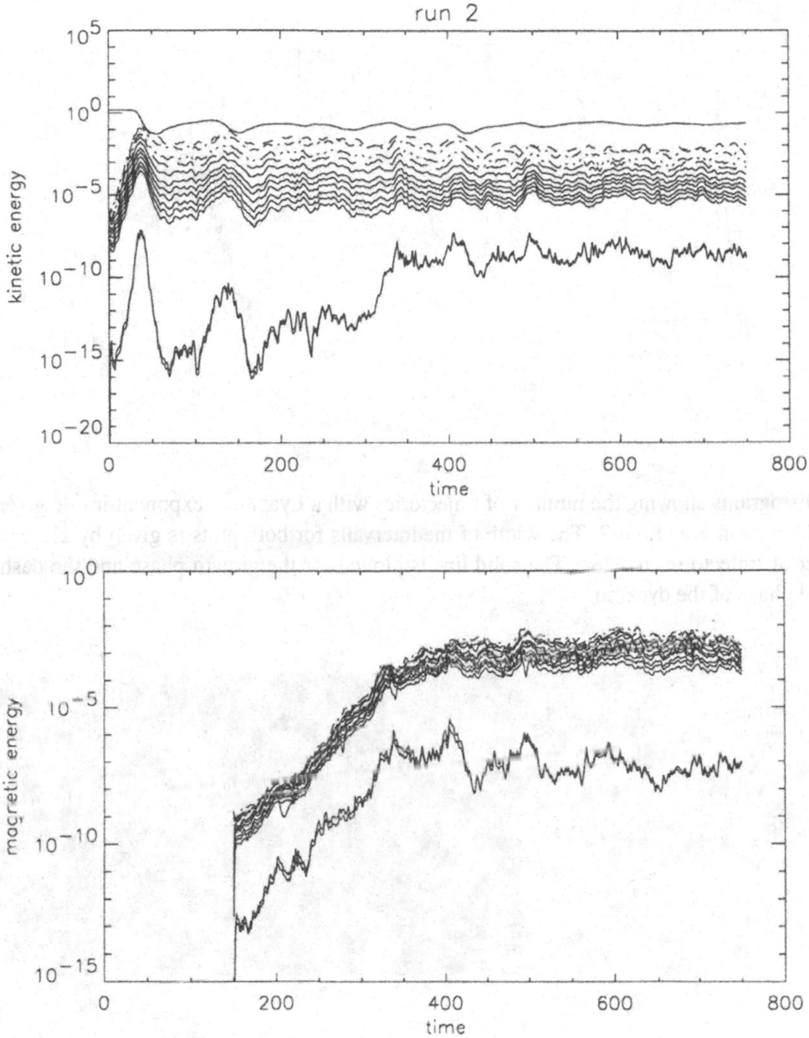

Figure 1. Time evolution of kinetic and magnetic energies in different shells of k-vectors for run 2. The solid line with the highest kinetic energy corresponds to the ABC_1 flow. It is hydrodynamically unstable and destabilizes without the influence of magetic field. At time 150 the magnetic seed field is introduced. It grows exponentially in the kinematic phase of the dynamo. At time 350 the nonlinear saturation of magnetic field begins. The two lowest curves in both plots correspond to shells 31 and 32.

flow is visualized in figure 3, where the greyscale images of the ABC-flow and the flow in the growth phase are opposed to each other. In the saturated phase, were the magnetic field energy now is 12% of the kinetic energy of the flow, chaos now clearly is diminished: the corresponding histogram in figure 2 is shifted to the left, and in figure 3 the corresponding grey scale image

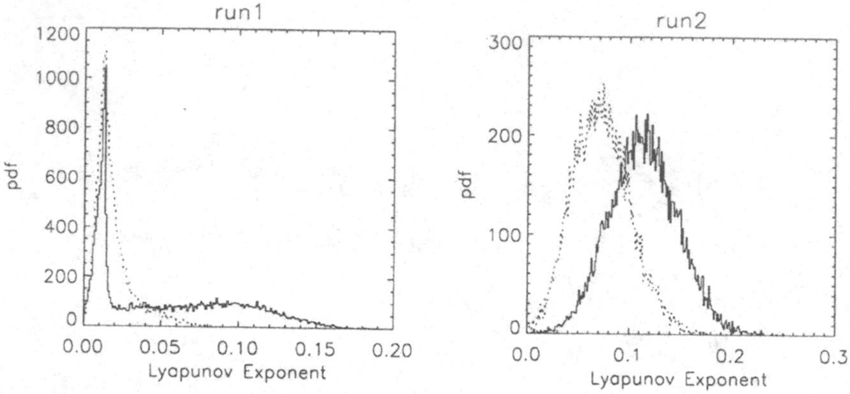

Figure 2. Histograms showing the number of trajectories with a Lyapunov exponent inside given intervalls of length $\Delta\lambda$ for run 1 and run 2. The width of the intervalls for both plots is given by $\Delta\lambda = 0.001$; the total number of trajectories is 128^2. The solid line is plotted for the growth phase and the dashed line for the saturated phase of the dynamo.

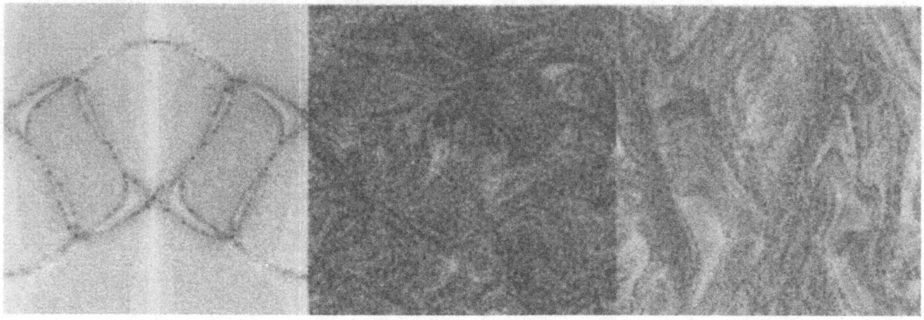

Figure 3. Grey scale images of finite–time Lyapunov exponents ranging from 0 (white) to 0.25 (black). The first image on the left side is computed for the ABC_1–flow and shown as comparison to the ABC-forced flow in the growth phase of the dynamo of run 2 (center). The image on the right side finally is computed for the saturated phase of run 2.

is visibly brighter than in the growth phase. The mean value of the Lyapunov exponent in the saturated phase is 0.0731, which is 64% of the growth phase.

In run 3 we kept the values of the Reynolds numbers but switched to scale separation $k_0 = 2$. A doubling of scale separation also demands a doubling in resolution in each space direction (from 64^3 in run 2 to 128^3 in this run). The level of saturation now is considerable increased, we reach equipartition of kinetic and magnetic energy: $\chi \approx 1.0$. Here also the strength of chaos in the flow of the growth phase of the dynamo is considerably higher than in the ABC_2-flow as becomes clear from figure 4. Now the backreaction of the magnetic field by the Lorentz-

Figure 4. Grey scale images of finite–time Lyapunov exponents ranging from 0 (white) to 0.35 (black). Again the ABC flow (now for scale separation $k_0 = 2$, left image), the flow of the growth phase (center) and of the saturation phase (right) are shown, now for run 3.

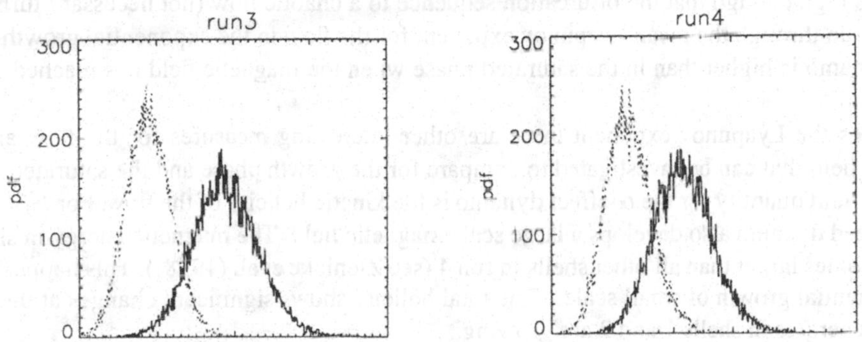

Figure 5. Histograms showing the number of trajectories with a Lyapunov exponent inside given intervalls of length $\Delta\lambda$ for run 3 and run 4. The width of the intervalls is given by $\Delta\lambda = 0.001$ for run 3 and by $\Delta\lambda = 0.0025$ for run 4. The solid line represents the growth phase and the dashed line represents the saturated phase of the dynamo.

force reaches its full strength, which shows itself in the data for the Lyapunov exponent: see the histogram in figure 5 (left hand side) and the pixel graphic in figure 4. The mean value of the Lyapunov exponents in the growth and saturated phases are 0.1892 respective 0.0901 showing a diminution to 48% in the saturated phase compared to the growth phase.

In run 4 the scale separation is increased to $k_0 = 4$, but the kinetic and magnetic Reynolds numbers are reduced to 12. Nevertheless, the ABC_4-flow already is hydrodynamically unstable and a more chaotic state is reached without magnetic field, before the seed field is introduced. A high value of the ratio of $\chi = E^M/E^V \approx 1.4$ is found. Also in this run we find a diminition of chaos from the growth phase to the saturated phase as can be seen from the histogram in figure

5 (right hand side) and table 1: the mean value of the LE in the saturated phase is 59% of that in the growth phase.

5 Conclusions

The investigation of the saturation process of a nonlinear dynamo is numerically challenging insofar as it is necessary to integrate the full MHD-equations in three-dimensional space. Besides the inertia term of the Navier-Stokes equation also the Lorentz force and the interaction between the velocity field and the magnetic field add nonlinear terms in the MHD-equations. All these nonlinearities must be included to check the hypothesis wether the backreaction of a growing magnetic field suppresses Lagrangian chaos of the underlying flow, which is the motor to amplify magnetic field at small scales in the fast dynamo. The magnetic Reynolds numbers reached in our calculations are still small compared with the magnetic Reynolds numbers of astronomical objects and thus we could not deal with the fast dynamo properly (fast dynamo means the limit $R^M \to \infty$). Nevertheless, the tendency from our results seems to be clear: If the kinetic Reynolds number is high enough that the bifurcation sequence to a chaotic flow (not necessary turbulent) has been run through the mean Lyapunov exponent for the flow in the exponential growth phase of the dynamo is higher than in the saturated phase when the magnetic field has reached its full strength.

Besides the Lyapunov exponent there are other interesting measures for the flow and the magnetic field that can be investigated to compare for the growth phase and the saturated phase. The important quantity for the α-effect dynamo is the kinetic helicity of the flow. For $k_0 = 4$ the ABC-forced dynamo also develops a large scale magnetic field. The magnetic energy in shells 1 and 2 becomes larger than all other shells in run 4 (see Zienicke et al. (1998)). This happens after the exponential growth of small scales. The total helicity shows significant changes at the times were the energies in shells 1 and 2 are growing.

Another very important question is the spatial structure of the flow as well as the magnetic field. A crude measure for the intermittency of the magnetic field can be defined by the ratio b_{max}/b_{rms}, where b_{max} is the maximum value of the magnetic field on all collocation points in the computational domain. b_{max}/b_{rms} changes by a factor between 1/3 and 1/4 between growth and saturated phase in run 3 (see Zienicke et al. (1998)). This is a hint that the spatial structure of the magnetic fields is really different from each other in the growth phase and in the saturated phase as suggested by Blackman (1996). These issues should be further investigated in the future.

Acknowledgements: Computations were done on T3E/IDRIS (Orsay). We are pleased to acknowledge financial support from CNRS–1202–MFGA & EEC–ERBCHRXCT930410.

References

V.I. Arnold (1965), Sur la topologie des écoulements stationnaires des fluides parfaits, C. R. Acad. Sci. Paris **261**, 17 - 20.

V.I. Arnold and E.I. Korkina (1983). The Growth of a Magnetic Field in a Threedimensional Steady Incompressible Flow, Vest. Mosk. Un. Ta. Ser. 1, Matem. Mekh., no. 3, 43 - 46.

D. Balsara and A. Pouquet, Physics of Plasmas, to appear

A. Bhattacharjee and Y. Yuan (1995). Astrophys. Journ. **449**, 739B.

E. Blackman (1996). Overcoming the Backreaction on Turbulent Motions in the Presence of Magnetic Fields, Phys. Rev. Lett. **77**, 2694 - 2697.

A. Brandenburg et al. (1996). Magnetic structures in a Dynamo Simulation, Journ. Fluid Mech. **306**, 325-352.

C. Canuto, M. Hussaini, A. Quarteroni, T. Zang (1988). Spectral Methods in Fluid Dynamics, Springer, Berlin.

F. Cattaneo, K. Hughes, E. Kim (1996). Suppression of Chaos in a Simplified Nonlinear Dynamo Model, Phys. Rev. Lett. **76**, 2057 - 2060.

S. Childress (1970). New Solutions of the Kinematic Dynamo Problem, Journ. Math. Phys. **11**, 3063 - 3076.

S. Childress and A.D. Gilbert (1995). Stretch, Twist, Fold: The Fast Dynamo, Springer, Berlin.

T. Dombre et al. (1986). Chaotic Streamlines in the ABC Flows, Journ. Fluid. Mech. **167**, 353 - 391.

F. Feudel, N. Seehafer, O. Schmidtmann (1996a). Bifurcation Phenomena of the Magnetofluid Equations, Math. and Comp. in Sim. **40**, 235 - 245.

F. Feudel, N. Seehafer, B. Galanti, S. Rüdiger (1996b). Symmetry-Breaking Bifurcations for the Mageto-hydrodynamic Equations with Helical Forcing, Phys. Rev. E **54**, 2589-2596.

U. Frisch, A. Pouquet, J. Leorat, A. Mazure (1975). Possibility of an inverse cascade of magnetic helicity in magnetohydrodynamic turbulence, Journ. Fluid Mech. **68**, 769-778.

D.J. Galloway and U. Frisch (1984). A Numerical Investigation of Magnetic Field Generation in a Flow with Chaotic Streamlines, Geophys. Astrophys. Fluid Dyn. **29**, 13 - 18.

D.J. Galloway and U. Frisch (1986). Dynamo Action in a Family of Flows with Chaotic Streamlines, Geophys. Astrophys. Fluid Dyn. **36**, 53 - 83.

B. Galanti, P.L. Sulem, A. Pouquet (1992). Linear and Non-linear Dynamos Associated with ABC Flows, Geophys. Astrophys. Fluid Dyn. **66**, 183 - 208.

D. Gottlieb and S. Orzag (1977). Numerical Analysis of Spectral Methods: Theory and Applications, SIAM-CBMS, Philadelphia.

J. Guckenheimer and P. Holmes (1983). Nonlinear Oscillatons, Dynamical Systems and Bifurcations of Vector Fields, Springer, New York.

M. Hénon (1966). Sur la topologie des lignes de courant dans un cas particulier, C. R. Acad. Sci. Paris A **262**, 312 - 314.

R. Horiuchi and T. Sato (1988). Phys. Fluids **31**, 1142.

F. Krause and K.-H. Rädler (1980). Mean Field Magnetohydrodynamics and Dynamo Theory, Akademie Verlag, Berlin.

Y.-T. Lau and J.M. Finn (1993). Fast Dynamos with Finite Resistivity in Steady Flows with Stagnation Points, Phys. Fluids B **5**, 365 - 375.

M. Meneguzzi, U. Frisch, A. Pouquet (1981). Helical and nonhelical turbulent dynamos, Phys. Rev. Let. **47**, 1060-1064.

H.K. Moffatt (1978). Magnetic Field Generation in Electrically Conducting Fluids, Cambridge University Press, Cambridge.

E. Ott (1993). Chaos in Dynamical Sustems, Cambridge University Press, Cambridge.

O. Podvigina and A. Pouquet (1994). On the Non-linear Stability of the 1:1:1 ABC Flow, Physica D **75**, 471 - 508

A. Pouquet, U. Frisch, J. Léorat (1976). Strong MHD helical turbulence and the nonlinear dynamo effect, Journ. Fluid. Mech. **77**, 321-354.

A. Pouquet and G.S. Patterson (1978). Numerical simulation of helical magnetohydrodynamic turbulence, Journ. Fluid Mech. **85**, 305-323.

P.H. Roberts and A.M. Sowards (1992). Dynamo Theory, Ann. Rev. Fluid Mech. **24**, 459-512.

E. Zienicke, H. Politano, A. Pouquet (1998). Variable Intensity of Lagrangian Chaos in the Nonlinear Dynamo Problem, Phys. Rev. Lett. **81**, 4640 - 4643.

A. Brandenburg et al. (19[illegible]), Astrophys. Journ. Lett. [illegible]

[illegible] (199[illegible]) Evidence for the [illegible] Motions in the Presence of Magnetic Field, Phys. Rev. Lett. 7[illegible], 26[illegible]-2651.

A. Brandenburg et al. (1996), Magnetic Structures in a Dynamo Simulation, Journ. Fluid Mech. 306, 325-352.

C. Canuto, M. Hussaini, A. Quarteroni, T. Zang (1988) Spectral Methods in Fluid Dynamics, Springer, Berlin.

F. Cattaneo, D. Hughes, E. Kim (1996) Suppression of Chaos in a Simplified Nonlinear Dynamo Model, Phys. Rev. Lett. 76, 2057-2060.

S. Childress (1970), New Solutions of the Kinematic Dynamo Problem, Journ. Math. Phys. 11, 3063-3076.

S. Childress and A.D. Gilbert (1995) Stretch, Twist, Fold: The Fast Dynamo, Springer, Berlin.

T. Dombre et al. (1986) Chaotic Streamlines in the ABC Flows, Journ. Fluid Mech. 167, 353-391.

F. Feudel, N. Seehafer, O. Schmidtmann (1996a) Bifurcation Phenomena of the Magnetofluid Equations, Math. and Comp. in Simul. 40, 235-245.

F. Feudel, N. Seehafer, B. Galanti, S. Rüdiger (1996b) Symmetry-breaking Bifurcations for the Magnetohydrodynamic Equations with Helical Forcing, Phys. Rev. E 54, 2589-2596.

U. Frisch, A. Pouquet, J. Léorat, A. Mazure (1975) Possibility of an inverse cascade of magnetic helicity in magnetohydrodynamic turbulence, Journ. Fluid Mech. 68, 769-778.

D.J. Galloway and U. Frisch (1984) A Numerical Investigation of Magnetic Field Generation in a Flow with Chaotic Streamlines, Geophys. Astrophys. Fluid Dyn. 29, 13-18.

D.J. Galloway and U. Frisch (1986) Dynamo Action in a Family of Flows with Chaotic Streamlines, Geophys. Astrophys. Fluid Dyn. 36, 53-83.

B. Galanti, P.L. Sulem, A. Pouquet (1992) Linear and Non-linear Dynamos Associated with ABC Flows, Geophys. Astrophys. Fluid Dyn. 66, 183-208.

D. Gottlieb and S. Orszag (1977) Numerical Analysis of Spectral Methods: Theory and Applications, SIAM-CBMS, Philadelphia.

J. Guckenheimer and P. Holmes (1983) Nonlinear Oscillations, Dynamical Systems, and Bifurcations of Vector Fields, Springer, New York.

M. Hénon (1966) Sur la topologie des lignes de courant dans un cas particulier, C. R. Acad. Sci. Paris A 262, 312-314.

K. [illegible] and T. Sato (19[illegible]) Phys. Fluids 5[illegible], [illegible].

F. Krause and K.-H. Rädler (1980) Mean-Field Magnetohydrodynamics and Dynamo Theory, Akademie-Verlag, Berlin.

Y.-T. Lau and J.M. Finn (1993) Fast Dynamos with Finite Resistivity in Steady Flows with Stagnation Points, Phys. Fluids B 5, 365-375.

M. Meneguzzi, U. Frisch, A. Pouquet (1981) Helical and nonhelical turbulent dynamos, Phys. Rev. Lett. 47, 1060-1064.

H.K. Moffatt (1978) Magnetic Field Generation in Electrically Conducting Fluids, Cambridge University Press, Cambridge.

E. Ott (1993) Chaos in Dynamical Systems, Cambridge University Press, Cambridge.

O. Podvigina and A. Pouquet (1994) On the non-linear stability of the 1:1:1 ABC Flow, Physica D 75, 471-508.

A. Pouquet, U. Frisch, J. Léorat (1976) Strong MHD helical turbulence and the nonlinear dynamo effect, Journ. Fluid Mech. 77, 321-354.

A. Pouquet and G.S. Patterson (1978) Numerical simulation of helical magnetohydrodynamic turbulence, Journ. Fluid Mech. 85, 305-323.

[illegible] Roberts [illegible] (19[illegible]) Dynamo [illegible]

[illegible] (199[illegible]) Variational [illegible] Problem, Phys. Rev. Lett. [illegible]

Chapter VI

Ch. Karcher, U. Lüdtke, D. Schulze, A. Thess

COMPUTATIONAL MHD
PART III -
ELECTROMAGNETIC CONTROL OF CONVECTIVE FLOWS

Computational MHD
Part III – Application to Electromagnetic Control of Convective Flows

Christian Karcher[1], Ulrich Lüdtke[2], Dietmar Schulze[2], and Andre Thess[1]

[1] Department of Mechanical Engineering, Ilmenau University of Technology, Germany
[2] Department of Electrical Engineering, Ilmenau University of Technology, Germany

Abstract: Convective flow in a liquid metal heated locally at its upper surface and affected by an applied time-dependent magnetic field is investigated. The system under consideration serves as a physical model for the industrial process of electron beam evaporation of liquid metals. In this process, the strong energy input induces strong temperature gradients along the free surface and in the interior of the melt. Thus, the liquid metal is subject to both thermocapillary and natural convection. The vigorous convective motion within the melt leads to highly unwelcome heat losses through the walls of the crucible. The strong convective heat transfer limits the temperature rise in the hot spot and, therefore, the thermodynamic efficiency of the evaporation process. The present paper aims to demonstrate that the melt-flow can be effectively controlled by using external magnetic fields in order to considerably reduce the convective heat losses. As examples, we employ numerical simulations based on the finite element method to study the effects of both a traveling magnetic field and a rotating magnetic field.

1 Introduction

Electron beam evaporation of liquid metals (Schiller [1]) is an innovative PVD-technology increasingly used in industrial application to produce high-quality coatings. In this vacuum process a high-energy electron beam bombards the surface of a metal ingot. At the surface, the kinetic energy of the electrons is transformed into heating power. The ingot melts and forms a free surface. When the temperature of the free surface exceeds the actual (pressure-dependent) saturation temperature of the ambient gas, the liquid starts to vaporize. The rising vapor cloud condenses on a moving substrate located at some distance above. A sketch of the evaporation process is shown in Figure 1. To guarantee a superior quality of the coating, the melt is typically confined in a water-cooled copper crucible. The intensive cooling prevents chemical reactions between the melt and the crucible walls. By that, crucible materials and their reaction products are practically excluded from evaporation.

A drawback of the process is that only a small portion (< 3%) of the electron beam power is converted into vapor energy. This poor profit results from the fact that the strong energy flux from the electron beam induces strong temperature gradients along the free surface and in the interior of the melt. Hence, the liquid metal is subject to both surface-tension-driven (Davis [2]) and buoyancy-driven (Siggia [3]) convection. The strong convective heat transfer limits the temperature rise at the free surface and therefore the thermodynamic efficiency of the

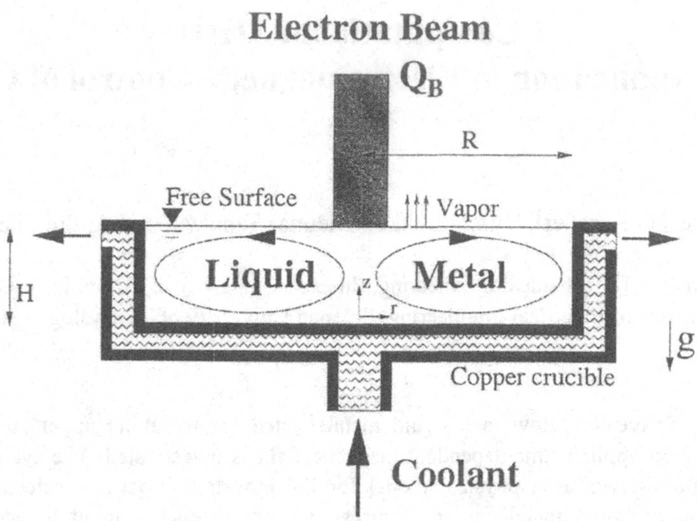

Figure 1: Sketch of the electron beam evaporation process.

evaporation process, see Refs. [4]-[7]. In typical applications these highly unwelcome heat losses amount for up to 70% of the power input.

Our goal is to reduce the conductive heat losses during electron beam evaporation by controlling convection using an applied magnetic field. Furthermore, with this practical example we aim to demonstrate how a coupled electromagnetic and hydrodynamic problem, governed by both the Maxwell and Navier Stokes equations, can be treated numerically. Here we exploit the physical mechanism that in electrically conducting fluids, interactions between the applied magnetic field and the fluid-flow generate Lorentz forces (cf. Moreau [8]). This body force may serve to slow down the flow or to superimpose a secondary fluid motion that counteracts the convective transport. The effects of a static vertical magnetic field on heat transfer in a liquid locally heated at its upper surface are discussed in Refs. [9]- [11]. In this case the Lorentz forces acting on the melt can be obtained analytically. It turns out that the uniform field tends to dampen convective motion. The predictions of numerical simulations of the fluid dynamics are in good agreement with observations in model experiments [11]. However, relatively strong uniform magnetic fluxes ($B_0 \approx 150$mT) are necessary to achieve the desired reduction of the convective heat losses. In the present study we focus on the effect of both a traveling magnetic field and a rotating magnetic field on surface-tension-driven convection in a liquid metal heated locally from above. In this case it is expected that even relatively weak magnetic fluxes ($B_0 \approx 5$mT) can reduce the convective heat losses. Here the resulting Lorentz forces generally cannot be calculated analytically but have to be derived numerically by a separate electromagnetic simulation.

Therefore, our numerical approach consists of two steps. We first solve the electromagnetic equations for a given geometry of both inductor and crucible and a prescribed inductor voltage. As a basic finding we obtain the magnetic flux lines in the melt and the distribution of the Lorentz forces acting on the liquid metal. In a second step, we solve the hydrodynamic equations describing the convective heat transfer within the melt. As an input we use here the time-averaged distribution of the Lorentz forces obtained by the electromagnetic simulations. The present approach neglects the full coupling between the electrodynamics and the fluid dynamics of the problem, i.e. that the melt-flow affects the magnetic field. This simplification is justified for small magnetic Reynolds numbers, see eg. Moreau [8].

The present study is organized as follows. Sec. 2 describes the numerical method used in the electromagnetic simulations. We show magnetic field lines and distributions of Lorentz forces for the cases that the inductor generates a traveling and a rotating magnetic field. In Sec. 3 we describe the fluid dynamics of the problem and show how the melt-flow is affected by the applied magnetic fields. Finally, in Sec. 4 we briefly summarize the main findings.

2 Electromagnetic Simulations

2.1 Basic Equations

The starting point in the mathematical description of electromagnetic problems are the Maxwell equations together with the so-called material equations, see eg. Binns et al. [12] and Jackson [13]. These equations read as

$$\nabla\times\mathbf{H}=\mathbf{J}\,,\ \nabla\times\mathbf{E}=-\frac{\partial}{\partial t}\mathbf{B}\,,\ \nabla\cdot\mathbf{B}=0\,,\ \mathbf{J}=\sigma\mathbf{E}\,,\ \mathbf{B}=\mu\mathbf{H}\,. \tag{1a-e}$$

Here, **H** and **B** denote the magnetic field and the magnetic flux, respectively, while **E** and **J** are the electric field and the electric current density. The material properties are the electrical conductivity $\sigma(\mathbf{x})$ and the permeability $\mu(\mathbf{x})$, which are both functions of the spatial position **x**. Note that in Eq. (1a) we have neglected the displacement flux. Moreover, in Ohm's law (Eq. 3(d)) we have neglected the induced current due to the interactions between the liquid metal flow and the applied magnetic field. In this so-called small magnetic Reynolds number limit, the electromagnetic problem decouples from the fluid dynamics [8], as already mentioned above.

It is convenient to replace the solenoidal field **B**, cf. Eq. (1c), by the magnetic vector potential **A**. We define

$$\mathbf{B}=\nabla\times\mathbf{A}\,. \tag{2}$$

Upon integrating Eq. (1b) and using Eq. (1d) we find the following equations for **E** and **J**:

$$\mathbf{E}=-\frac{\partial}{\partial t}\mathbf{A}-\nabla\Phi_0,\ \mathbf{J}=-\sigma\frac{\partial}{\partial t}\mathbf{A}-\sigma\nabla\Phi_0\,, \tag{3a,b}$$

where Φ_0 denotes the imposed inductor voltage. Finally, the magnetic vector potential **A** is governed by the equation

$$\nabla\times\left(\frac{1}{\mu}\nabla\times\mathbf{A}\right)+\sigma\frac{\partial}{\partial t}\mathbf{A}=-\sigma\nabla\Phi_0\,. \tag{4}$$

Since Eq. (4) defines an open-field problem, the boundary conditions for **A** take the form

$$\mathbf{A}=0 \text{ as } |\mathbf{x}|\to\infty\,. \tag{5}$$

Once we have found a solution to Eqs. (4) and (5), the Lorentz forces per unit mass, $\mathbf{F}_L$, acting on the liquid metal can be obtained by evaluating the relation

$$\mathbf{F}_L=\frac{1}{\rho}(\mathbf{J}\times\mathbf{B}), \tag{6}$$

where **J** and **B** are defined by Eqs. (3a) and (2), respectively, and ρ is the fluid density.

2.2 Numerical Method

For the electromagnetic simulations we use the finite-element code PROMETHEUS [14]. This home-made code allows solving the equation for the magnetic vector potential **A** (Eq. (4)) for any geometric arrangement and any materials. However, since the computational domain is limited, the far-field boundary condition (5) has to be replaced by a condition at finite distance. A typical choice is

$$\mathbf{A}=0 \text{ at } |\mathbf{x}|=5\cdot L\,, \tag{7}$$

where L is a characteristic length of the electromagnetic problem. In the present study L is the diameter of the crucible, i.e. L = 2R. The actual computations are carried out in the complex domain. Here, we assume that all quantities show a sinusoidal time-dependence. In this case the time derivative in Eq. (5) simplifies to

$$\frac{\partial}{\partial t}\underline{\mathbf{A}}=j\omega_0\underline{\mathbf{A}}\,, \tag{8}$$

where j is the imaginary unit, ω_0 is the frequency of the imposed inductor voltage Φ_0, and the bar denotes complex quantities.

PROMETHEUS provides a semi-automatic grid generation using macro-elements [15]. These elements are meshed automatically into linear finite elements and then put together for the total discretization of the computational domain. The resulting set of linear algebraic equations is solved iteratively using the incomplete Cholesky decomposition conjugate gradient (ICCG) method, see Press et al. [16] for details.

In the following we present two examples for an electromagnetic simulation, namely the generation of a traveling and of a rotating magnetic field by an inductor subjected to an imposed sinusoidal voltage. The magnitude of the inductor voltage is chosen so that the magnitude of the magnetic flux does not exceed B = 5mT in the region where the electron beam impinges on the liquid metal surface. This limitation guarantees that the beam is compatible with the magnetic field [17]. In both examples the computational effort can be reduced considerably since the problems remain two-dimensional and only one component of the magnetic vector potential is present.

2.3 Numerical Simulation of a Traveling Magnetic Field

Fig. 2 shows an axisymmetric arrangement for the generation of a traveling magnetic field together with a part of the mesh used in the numerical simulations. Here, three inductors (R, S, T), placed at the side wall of the crucible, are subjected to a three-phase sinusoidal voltage with a phase shift of 120° and a frequency of 16Hz. The crucible has an inner height of 70mm and an inner radius of 82.5mm while the wall thickness is 10mm. The inductor and crucible material is copper with an electrical conductivity of $\sigma = 5.6 \times 10^7$ S/m and a relative permeability of $\mu_r = 1$. The crucible is filled with liquid steel ($\sigma = 1.0 \times 10^6$ S/m, $\mu_r = 1$) to a height of H = 60mm. The surrounding material is air ($\sigma = 0$, $\mu_r = 1$). The entire mesh consists of about 10500 square elements.

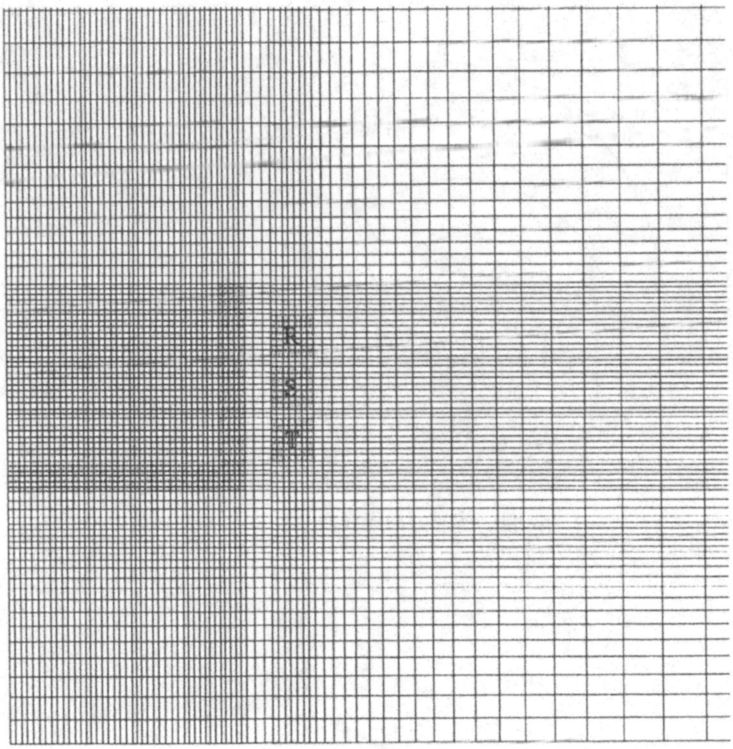

Figure 2: Arrangement to generate a traveling magnetic field and part of the computational mesh.

In the actual example, Eq. 4 has to be solved only for the azimuthal component of the magnetic vector potential, i.e.

$$\underline{\mathbf{A}} = \underline{A}_{\varphi}(r,z) \cdot \mathbf{e}_{\varphi} \,. \tag{9}$$

The induced current density and the induced magnetic are then calculated from Eqs. (3b) and (2), respectively. We obtain

$$\underline{\mathbf{J}} = \underline{J}_{\varphi}(r,z) \cdot \mathbf{e}_{\varphi} \,, \; \underline{\mathbf{B}} = \underline{B}_{r}(r,z) \cdot \mathbf{e}_{r} + \underline{B}_{z}(r,z) \cdot \mathbf{e}_{z} \,. \tag{10a,b}$$

Fig. 3 shows a snapshot of the calculated magnetic flux lines for the case that the phase of inductor S is zero. The flux lines travel periodically in vertical direction. As obvious, the flux lines penetrate through the crucible wall into the melt. Finally, Fig. 4 shows the distribution of the resulting time-averaged Lorentz forces $F_r(r,z)$ and $F_z(r,z)$ acting on the melt, as calculated from Eq. (6). They mainly act near the side wall and tend to push fluid towards the free surface.

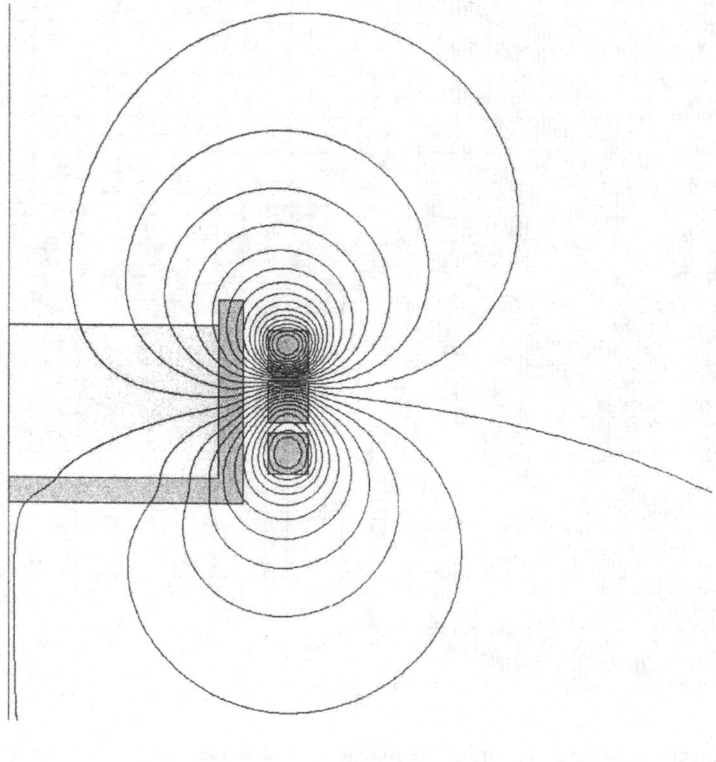

Figure 3: Snapshot of the magnetic flux lines.

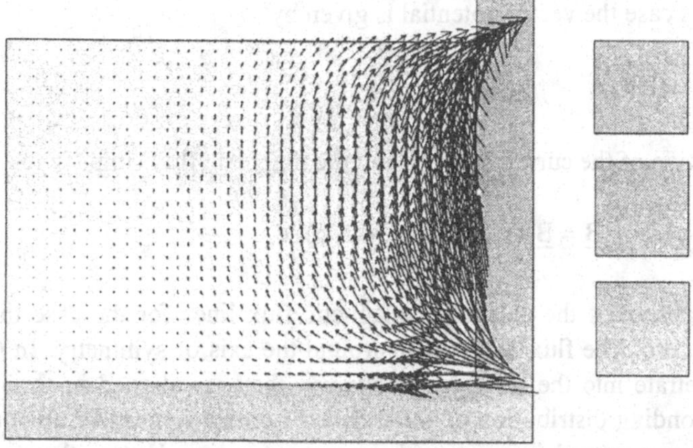

Figure 4: Time-averaged distribution of the Lorentz forces.

2.4 Numerical Simulation of a Rotating Magnetic Field

Fig. 5 shows an arrangement for the generation of a rotating magnetic field together with a part of the computational mesh. In the present case the mesh is built up by both triangular and square elements. Here, three pairs of inductors (R,-R), (S,-S), (T,-T) are placed along the circumference of the crucible. Again, they are fed by a three-phase voltage with a phase shift of 120° and a frequency of 16Hz. The material parameters are the same as in the previous example.

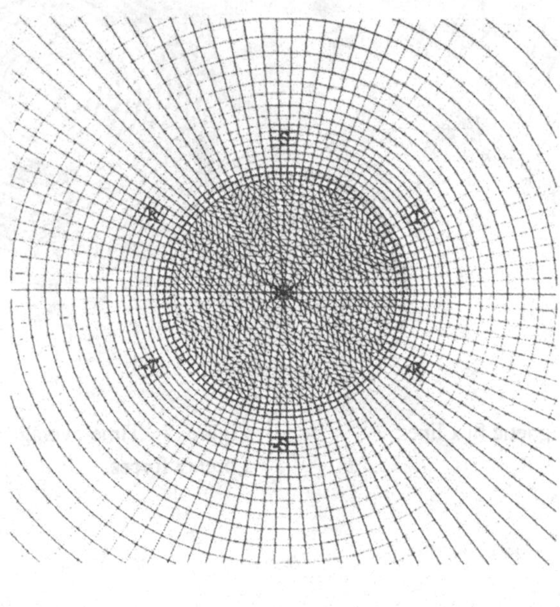

Figure 5: Arrangement to generate a rotating magnetic field and part of the computational mesh.

For simplicity, the problem is considered to be independent of the axial coordinate ($\partial / \partial z \equiv 0$). In this case the vector potential is given by

$$\underline{\mathbf{A}} = \underline{A}_z(r,\varphi) \cdot \mathbf{e}_z , \tag{11}$$

and the representation of the current density and the magnetic flux simplify to

$$\underline{\mathbf{J}} = \underline{J}_z(r,\varphi) \cdot \mathbf{e}_z , \quad \underline{\mathbf{B}} = \underline{B}_r(r,\varphi) \cdot \mathbf{e}_r + \underline{B}_\varphi(r,\varphi) \cdot \mathbf{e}_\varphi . \tag{12a,b}$$

Fig. 6 shows a snapshot of the calculated magnetic flux lines for the case that the phase of inductor (R,-R) is zero. The flux lines rotate around the axis of symmetry. In the present case the flux lines penetrate into the melt, too. However, they are aligned in straight lines. Fig. 7 shows the corresponding distribution of the induced Lorentz forces. As obvious, they mainly act in azimuthal direction and increase with increasing radius. Hence, the generated rotating magnetic field tends to stir the melt.

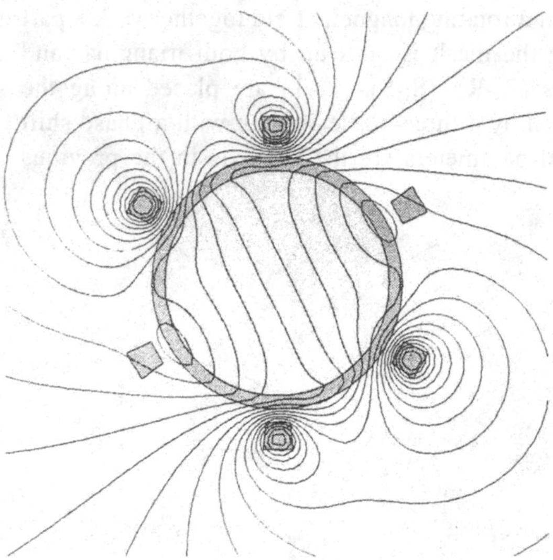

Figure 6: Snapshot of the magnetic flux lines.

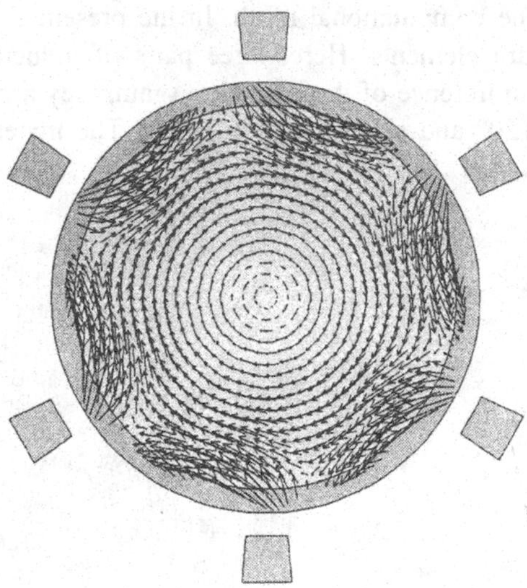

Figure 7: Time-averaged distribution of the Lorentz forces.

3 Fluid Dynamic Simulations

3.1 Basic Equations and Parameters

For the numerical simulations we consider the Boussinesq equations (cf. Platten and Legros [18]) in a cylindrical (r-φ-z) coordinate system, cf. Fig. 1. The governing equations are formulated in the streamfunction-vorticity-temperature (ψ-ω-T) representation. Furthermore, we solve an additional transport equation for the angular momentum Γ of the melt-flow. This equation and the vorticity equation are equipped with additional terms that account for the Lorentz forces induced by a rotating and traveling magnetic field, respectively. For simplicity, we assume that the material in the crucible is all liquid, that the flow is axisymmetric $(\partial/\partial\varphi \equiv 0)$, that the effects of buoyancy and Joule heating can be neglected and flow is laminar. Within these assumptions the set of governing equations writes as follows. In dimensionless form we obtain

$$\left(\frac{\partial^2}{\partial r^2} - \frac{1}{r}\frac{\partial}{\partial r} + \frac{\partial^2}{\partial z^2}\right)\psi = r\omega . \tag{13}$$

$$\left(\frac{\partial}{\partial t} + \frac{1}{r}\frac{\partial\psi}{\partial z}\frac{\partial}{\partial r} - \frac{1}{r}\frac{\partial\psi}{\partial r}\frac{\partial}{\partial z} - \frac{1}{r^2}\frac{\partial\psi}{\partial z}\right)\omega = \Pr\left(\frac{\partial^2}{\partial r^2} + \frac{1}{r}\frac{\partial}{\partial r} - \frac{1}{r^2} + \frac{\partial^2}{\partial z^2}\right)\omega$$
$$+ \frac{2}{r^3}\Gamma\frac{\partial}{\partial z}\Gamma - \frac{\Pr}{Ma}Ha^2F_0\left(\frac{\partial}{\partial z}f_r - \frac{\partial}{\partial r}f_z\right). \tag{14}$$

$$\left(\frac{\partial}{\partial t} + \frac{1}{r}\frac{\partial\psi}{\partial z}\frac{\partial}{\partial r} - \frac{1}{r}\frac{\partial\psi}{\partial r}\frac{\partial}{\partial z}\right)T = \left(\frac{\partial^2}{\partial r^2} + \frac{1}{r}\frac{\partial}{\partial r} + \frac{\partial^2}{\partial z^2}\right)T . \tag{15}$$

$$\left(\frac{\partial}{\partial t} + \frac{1}{r}\frac{\partial\psi}{\partial z}\frac{\partial}{\partial r} - \frac{1}{r}\frac{\partial\psi}{\partial r}\frac{\partial}{\partial z} + \frac{1}{r^2}\frac{\partial\psi}{\partial z}\right)\Gamma = \frac{\Pr}{Ma}\left(\frac{\partial^2}{\partial r^2} + \frac{1}{r}\frac{\partial}{\partial r} - \frac{1}{r^2} + \frac{\partial^2}{\partial z^2}\right)\Gamma$$
$$+ r\frac{\Pr}{Ma}Ha^2G_0 \cdot g_\varphi . \tag{16}$$

Here, the vorticity ω, the streamfunction ψ, and the angular momentum Γ are defined as

$$\omega = \frac{\partial}{\partial z}u - \frac{\partial}{\partial r}w,\ u = \frac{1}{r}\frac{\partial}{\partial z}\psi,\ w = -\frac{1}{r}\frac{\partial}{\partial r}\psi\ ,\ \Gamma = v \cdot r\ , \tag{17a-d}$$

where (u,v,w) denote the components of the velocity vector. In Eqs. (14), (16) the parameters F_0 and G_0 characterize the order of magnitude of the Lorentz forces induced by the traveling and rotating magnetic field, respectively, while $f_r(r,z)$, $f_z(r,z)$, and $g_\varphi(r)$ are functions of the order unity that represent the spatial distribution of the respective components of the Lorentz forces, cf.

Figs. (4), (7). The dimensionless groups in the governing equations (13)-(16) are the Prandtl number Pr, the Marangoni number, and the Hartmann number Ha. These parameters are given by

$$\mathrm{Pr}=\frac{\upsilon}{\kappa},\quad \mathrm{Ma}=\frac{\gamma Q}{\rho\nu\kappa\lambda},\quad \mathrm{Ha}=B_0 H\left(\frac{\sigma}{\rho\nu}\right)^{1/2}. \qquad \text{(18a-c)}$$

Here, the constant material properties of the liquid metal are the viscosity ν, the thermal diffusivity κ, the temperature coefficient of surface tension γ, the density ρ, the thermal conductivity λ, and the electrical conductivity σ, while Q and B_0 denote the energy input and the characteristic magnetic induction, respectively. To derive the equations above we used the scales

$$r,z \propto H,\ t \propto \frac{H^2}{\kappa \mathrm{Ma}},\ T-T_0 \propto \frac{Q}{\lambda H},\ \omega \propto \frac{\kappa \mathrm{Ma}}{H^2},\ \psi \propto \kappa \mathrm{Ma} H,\ \Gamma \propto \kappa \mathrm{Ma}. \qquad (19)$$

The governing equations are supplemented by appropriate boundary conditions. We apply fixed temperature and fixed heat flux conditions at the crucible walls and at the upper surface, respectively. Moreover, the liquid metal obeys the no-slip condition at rigid boundaries. At the free surface we take into account that vorticity is generated due to the Marangoni effect. Furthermore, we use symmetry conditions along the line r = 0. For simplicity, we neglect the effects of surface deformation and the (secondary) contributions of thermal radiation and evaporation to the energy balance at the free surface. The boundary conditions used read as

$$\psi=\frac{\partial\Psi}{\partial z}=\Gamma=0 \text{ at } z=0,\ \psi=\frac{\partial\psi}{\partial r}=\Gamma=0 \text{ at } r=R/H. \qquad \text{(20a-f)}$$

$$\psi=\frac{\partial\Gamma}{\partial z}=0 \text{ and } \omega=-\frac{\partial T}{\partial r} \text{ at } z=1. \qquad \text{(21a-c)}$$

$$T=0 \text{ at } z=0 \text{ and } r=R/H,\ \frac{\partial}{\partial z}T=\exp\left[-\frac{r^2}{2r_H^2}\right] \text{ at } z=1. \qquad \text{(22a,b)}$$

$$\psi=\omega=\Gamma=\frac{\partial T}{\partial r}=0 \text{ at } r=0. \qquad \text{(23a-d)}$$

Here, we assume an exponential distribution of heating power at the free surface and r_H denotes the characteristic radius of the heated zone.

To quantify the convective heat transfer we introduce the Nusselt number Nu defined by

$$\mathrm{Nu}=\frac{T_{H,0}}{T_H}. \qquad (24)$$

Here, $T_H = T(r = 0, z = 1)$ denotes the actual temperature in the hot spot and the index 0 refers to the motionless state of pure heat conduction. When convection sets in, we have $T_H < T_{H,0}$ and $Nu > 1$. Thus, the Nusselt number characterizes the convective heat losses.

3.2 Numerical Method

The governing equations shown above are solved numerically with a finite element method. We use linear triangles for spatial representation and the classical backward Euler scheme for time discretization. The nonlinear terms are treated explicitly. Starting from an appropriate initial condition, the governing equations are integrated in time until a (quasi) steady state emerges. During the integration the new time step is successively adjusted to satisfy the CFL stability criterion (cf. Press et al. [16]). At each time step the resulting sets of linear algebraic equations are solved iteratively using a Gauss-Seidel method with simultaneous over-relaxation (SOR), Chebyshev acceleration, and odd/even ordering, see Ref. [16] for details of this numerical algorithm. The calculations were carried out on an equidistant grid of 161×161 elements. It turned out that this grid size is sufficient since the Nusselt number typically changes less than 1% upon doubling the number of elements in each direction.

A special feature of the ψ-ω formulation of fluid-flow problems is that a priori no boundary conditions for the vorticity ω are available at rigid walls, cf. Eq. (20). Therefore, at each time step the numerical solution of Eq. (14) must be performed iteratively, see eg. Anderson [19] for details. An initial guess of the value of ω at $z = 0$ and $r = R/H$ can be derived by evaluating Eq. (13) at the respective boundary. Using conditions (20a,b) and (20d,e) we obtain

$$\omega = -\frac{1}{r}\frac{\partial^2 \Psi}{\partial z^2} \text{ at } z = 0, \quad \omega = -\frac{1}{r}\frac{\partial^2 \Psi}{\partial r^2} \text{ at } r = R/H. \tag{25a,b}$$

Applying these boundary conditions together with conditions at the symmetry line (Eq. (23b)) and at the free surface (Eq. (21c)), it is easy to get a numerical solution to Eq. (14). We then solve Eq. (13) anew using the homogeneous boundary conditions (Eqs. (20a, 20d, 21a)) and the symmetry condition (Eq. (23a)) and repeat the iteration until the desired numerical accuracy is achieved. Typically this is the case after 10 iterations.

3.3 Effect of a Traveling Magnetic Field

We start our investigations with the case that a traveling magnetic field is present, cf. Fig. (3). This case is characterized by $F_0 \neq 0$ and $G_0 = 0$, cf. Eqs. (14,16). The Lorentz forces acting on the liquid metal are shown in Fig. (4). Physically, they act as a source of vorticity, cf. Eq. (14). On the other hand, in this case there is no source term for angular momentum, cf. Eq. (16). Any initial distribution of Γ gradually dies out due to viscous dissipation. Hence, during the simulations we can set $\Gamma \equiv 0$.

As an example, Fig. 8 shows results of a numerical simulation for the following set of parameters: $Ma = 0.5 \times 10^4$, $Pr = 0.01$, $Ha = 5$, $R/H = 1.375$, and $r_H/H = 0.25$. Fig. 8(a) shows the streamlines in absence of the traveling magnetic field, i.e. for the case $F_0 = 0$. The remaining graphs (Figs. 8(b-d)) highlight how this flow field is modified upon increasing the strength of the Lorentz force induced by the applied traveling magnetic field. In detail, Figs.8(b-d) show streamlines for the cases $F_0 = 0.01$, $F_0 = 0.1$, and $F_0 = 1$, respectively.

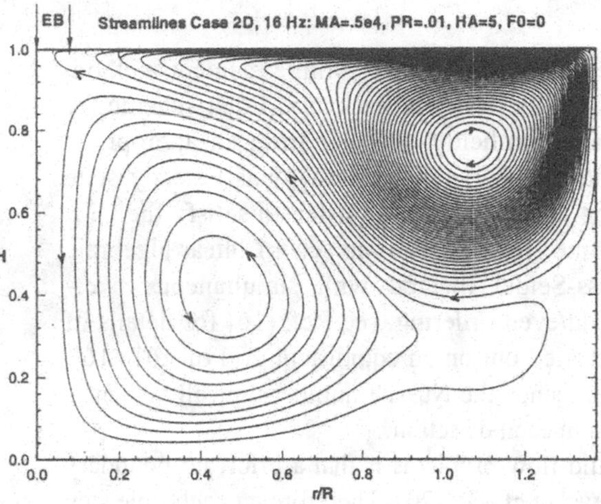

Figure 8(a): Streamlines for $F_0 = 0$.

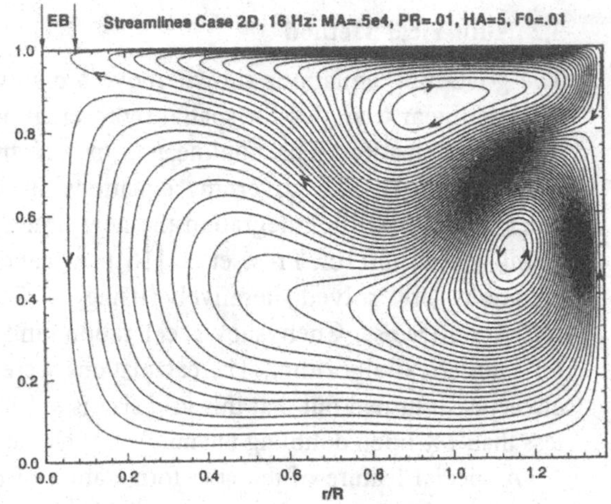

Figure 8(b): Streamlines for $F_0 = 0.01$.

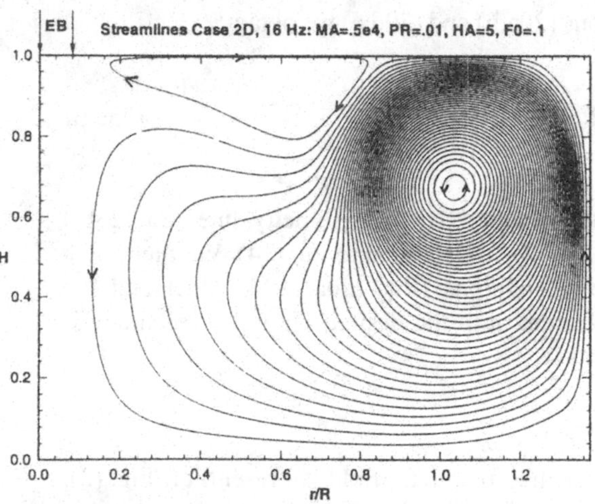

Figure 8(c): Streamlines for $F_0 = 0.1$.

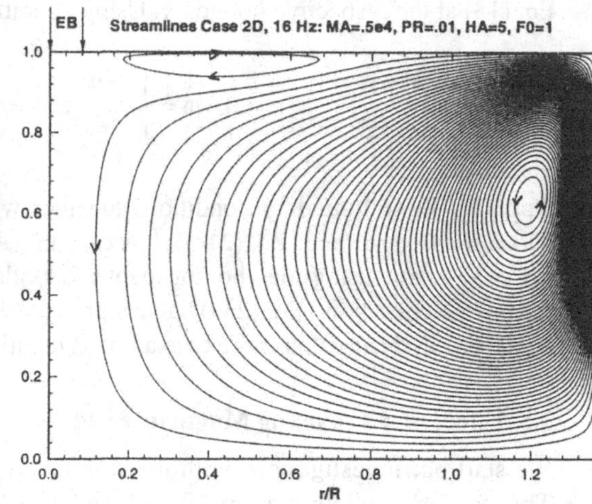

Figure 8(d): Streamlines for $F_0 = 1$

When the magnetic field is absent, the flow field consists of two global convection rolls of opposite circulation, see Fig. 8(a). Due to the Marangoni effect, a primary convection roll is induced at the free surface as fluid is pushed from the hot center towards the cold side wall. Here the liquid metal starts to sink, but separates from the wall at about half height. This separation effect creates the secondary convection roll in the lower part of the crucible.

On the other hand, when the traveling magnetic field is present the flow field is completely rearranged. For the case $F_0 = 0.01$ we observe that near the side wall a strong roll of counter-clockwise rotation starts to grow, cf. Fig. 8(b). This roll is clearly generated by the upward-pointing Lorentz forces in this region, cf. Fig. 4. By that, the primary convection roll shrinks considerably in size. Upon increasing the parameter G_0 up to $G_0 = 0.1$, the flow field induced by the Lorentz forces occupies almost the entire crucible, cf. Fig. 8(c). The Marangoni convection roll is displaced from the side wall. Hence, the radial heat transfer from the center to the wall is interrupted. Finally, for the case $G_0 = 1$, we observe that near the side wall a boundary layer-type flow is induced by the Lorentz forces, see Fig. 8(d). The convective flow due to the Marangoni effect is restricted to a small region near the heated surface.

In the example shown in Fig. 8(b), the actual reduction of the convective heat losses, however, typically amounts to just 2.5% compared to the case when the magnetic field is absent. We attribute this moderate reduction to the fact that the Marangoni numbers investigated in the numerical simulations are relatively small ($Ma = 0.5 \times 10^4$). In this parameter range the fluid-flow is very sensitive to the Lorentz forces. For instance, for the case $F_0 = 1$ the numerical simulations predict that the Nusselt number increases. Here the counter-clockwise fluid motion induced by the Lorentz forces enhances heat removal through the water-cooled bottom of the crucible. We conclude that the effect of a traveling magnetic field to reduce the convective heat losses become much more pronounced in parameter regimes that are typical in electron beam applications ($Ma \approx 1.0 \times 10^8$).

3.4 Effect of a Rotating Magnetic Field

We now turn to the case when a rotating magnetic field is present, cf. Fig. (6). This case is characterized by $G_0 \neq 0$ and $F_0 = 0$. The induced time-averaged Lorentz forces point in radial direction, as shown in Fig. (7). They act as a source of angular momentum, cf. Eq. (16), tending to stir the melt. Moreover, the generated rotation acts as a source of vorticity, cf. Eq. (14), especially in regions where the magnitude of $\partial\Gamma / \partial z$ is large. As obvious, this is the case adjacent to the bottom of the crucible where a so-called Ekman boundary layer (cf. Hopfinger [20]) is formed.

As an example, Fig. 9 shows results of a numerical simulation for the following set of parameters: $Ma = 0.5 \times 10^3$, $Pr = 0.01$, and $Ha = 5$, and the same geometry parameters as before. The plot of the streamlines in the left graph of Fig. 9 clearly demonstrates that under the action of a rotating magnetic field the flow field is completely rearranged. We observe two global convection rolls of opposite circulation which lie one on the top of the other. The centers of circulation are located near the cold wall where the induced Lorentz forces are the greatest, cf. Fig. 6. In the Ekman layer adjacent to the bottom, fluid is pushed towards the center. Moreover, we conclude that the magnetic field effectively counteracts the Marangoni effect as the radial flow induced by the Lorentz forces and surface-tension-driven convection, respectively, point in opposite direction. The right graph in Fig. 5 shows the distribution of the corresponding angular momentum Γ. We observe that in the bulk of the melt and at the free surface the liquid metal rotates like a rigid body. However, adjacent to the no-slip boundaries at the side wall and the bottom, the induced rotation dies out within thin boundary layers.

In the example shown in Fig. 9, the Nusselt number is $Nu = 1.0012$. This value indicates that convection is clearly suppressed by the rotating magnetic field. However, as for the case of a

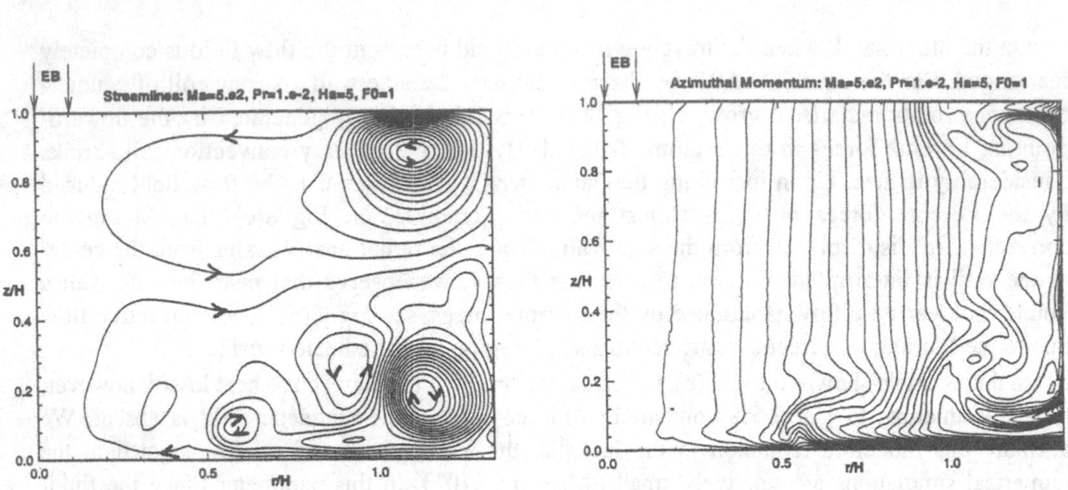

Fig. 9: Streamlines and angular momentum in the presence of a rotating magnetic field.

traveling magnetic field, the actual reduction of the convective heat losses is just about 2.5%. Again, this moderate reduction is because of the relatively low Marangoni numbers that are actually investigated in the numerical simulations.

4 Summary

We have investigated numerically the effects of both a traveling and a rotating magnetic field on surface-tension-driven convection in a liquid melt heated locally at its free surface. Due to the complexity of the problem we have split the numerical simulations into two steps. At the first, we perform an electromagnetic simulation to obtain the magnetic flux lines and resulting Lorentz forces acting on the melt. The second step is the simulation of the hydrodynamics of the melt affected by the induced Lorentz forces. The results show that the Lorentz forces generate a fluid motion that counteracts the Marangoni convection. However, the actual reduction of the convective heat losses was small. This is due to the relatively small Marangoni number of the flow. To study the effects at much higher Marangoni numbers typical in applications of electron beam technologies, would require incorporating into the simulations a fully three-dimensional model of the fluid dynamics, including turbulence modeling of free surface flows of liquid metals. This is a challenge for future research, see Thess and Schulze [21].

Acknowledgement: This work was sponsored by the Deutsche Forschungsgemeinschaft (DFG) under grant TH-C3.1.

References

1. S. Schiller, U. Heisig, S. Panzer: *Electron beam technology*, Technik Verlag Berlin (1982).
2. S.H. Davis: Ann. Rev. Fluid Mech. 19 (1987), 403.
3. E.D. Siggia: Ann. Rev. Fluid Mech. 26 (1994), 137.
4. A. Pumir, L. Blumenfeld: Phys. Rev. E54 (1997), 4528.
5. T. DebRoy, S.A. Davis: Rev. Mod. Phys. 67 (1995), 85.
6. Ch. Karcher, R. Schaller, Th. Boeck, Ch. Metzner, A. Thess: Int. J. Heat Mass Transfer 43 (2000), 1759.
7. Ch. Karcher, R. Schaller, A. Thess: in *Fluid flow phenomena in metals processing*, N. El-Kaddah, D.G.C. Robertson, S.T. Johansen, V.R. Voller (eds), TMS Publication (1999).
8. R.Moreau: *Magnetohydrodynamics*, Kluwer, Dordrecht (1990).
9. Ch. Karcher, Y. Kolesnikov, O. Andreev, A. Thess: Eur. J. Mech. B/Fluids (submitted).
10. Ch. Karcher, U. Lüdtke, D. Schulze, A. Thess: Proc. 3rd Int. Symposium on Electromagnetis Processing of Materials, Nagoya (2000), 473-478.
11. Ch. Karcher, Y. Kolesnikov, U. Lüdtke, A. Thess: Proc. 4th Pamir Int. Conference, Presqu'île Giens (2000), (to appear).
12. K. J. Binns, P. J. Lawrensen, C. W. Trowbridge: *The Analytical and Numerical Solution of Electric and Magnetic Fields*, Wiley & Sons, New York (1992).
13. D. J. Jackson: *Classical Electrodynamics*, J. Wiley and Sons, New York (1999).
14. U. Luedtke: Ilmenau University of Technology, PhD dissertation (1990).
15. O. C. Zienkiewicz, R. L. Taylor: *The Finite Element Method*, McGraw-Hill, London (1991).
16. W. H. Press, S. A. Teukolsky, W. T. Vetterling, B. B. Flannery: *Numerical Recipes in Fortran 77*, Cambridge University Press, New York (1992).
17. S. Rinke, Ch. Karcher, and A. Thess: Report ILR/B977, Dresden University of Technology (1998).
18. J. Platten, J. Legros: *Convection in Liquids*, Springer, New York (1984).
19. D. A. Anderson, J. C. Tannehill, R. H. Plechter: *Computational Fluid Mechanics and Heat Transfer*, Taylor and Francis, Washington D. C., (1997).
20. E. J. Hopfinger (Ed.): *Rotating Fluids in Geophysical and Industrial Applications*, CISM Courses and Lectures No. 329, Springer, New York (1992).
21. A. Thess, D. Schulze: This volume.

Printed in the United States
By Bookmasters

Printed in the United States
By Bookmasters